Springer Tracts in Autonomous Systems

Volume 1

The Springer Tracts in Autonomous Systems (STRAUS) publish the state-of-the-art and advances in unmanned systems exhibiting attributes of autonomy, intelligence, autonomous functionality and resilience. The objective is to provide the foundations of the overall cycle from design and modeling of unmanned systems, to control and implementation exhibiting autonomy and self-organization amidst extreme disturbances, also functioning in uncertain and dynamic environments.

The STRAUS series is the first of its kind in that it provides 'metrics' to evaluate autonomy and autonomous functionality and resilience, directly applicable to real complex engineering systems.

Books to be published are suitable for the scientist and engineer, practitioner and end-user. They are also suitable for graduate students and researchers who wish to conduct research in the topical areas covered by the series. Within the scope of the series are monographs, lecture notes, selected contributions from specialized conferences and workshops, as well as selected PhD theses.

More information about this series at http://www.springer.com/series/16447

Michail G. Michailidis · Kimon P. Valavanis ·
Matthew J. Rutherford

Nonlinear Control of Fixed-Wing UAVs with Time-Varying and Unstructured Uncertainties

Springer

Michail G. Michailidis
Ritchie School of Engineering and Computer
Science, Electrical and Computer
Engineering
University of Denver
Denver, CO, USA

Kimon P. Valavanis
Ritchie School of Engineering and Computer
Science, Electrical and Computer
Engineering
University of Denver
Denver, CO, USA

Matthew J. Rutherford
Ritchie School of Engineering and Computer
Science, Electrical and Computer
Engineering
University of Denver
Denver, CO, USA

ISSN 2662-5539 ISSN 2662-5547 (electronic)
Springer Tracts in Autonomous Systems
ISBN 978-3-030-40718-6 ISBN 978-3-030-40716-2 (eBook)
https://doi.org/10.1007/978-3-030-40716-2

This Springer imprint is published by the registered company Springer Nature Switzerland AG
The registered company address is: Gewerbestrasse 11, 6330 Cham, Switzerland

Go. Go far, over there, where the sun sinks into the endless sea. Lose yourself behind the horizon to face the unknown

—Philip Plisson, The Sea

Mpampa, Mama, Rena, this is for you!
(MGM)

To Stella and Panos, who have started their
own journey in life, and to Elissia who is
starting hers in college (KPV)

To my family (MJR)

Preface

Unmanned aircraft systems (UAS) and unmanned aviation, in general, has witnessed unprecedented levels of growth in the last decade, worldwide. Although military applications of unmanned aerial vehicles (UAVs) and UAS were dominant in the past, the map is changing, and it is postulated that civil and public domain applications will be on the forefront in the near future.

Despite initial research and development focus on traditional fixed-wing and helicopter UAV configurations, there have been major efforts towards designing and building quadrotors and other multi-rotor UAVs, bio-inspired UAVs, as well as other types of prototypes with enhanced aerodynamic capabilities. Such designs include morphing wing UAVs, circulation control based fixed-wing UAVs, segmented control surface UAVs, bird-looking UAVs, reconfigurable UAVs, among other "futuristic" designs.

However, when dealing with the so-called next generation, NextGen, UAVs, such systems have (fast) time-varying parameters and unstructured uncertainties; this is the case, for example, with circulation control based fixed-wing and morphing wing-based UAVs, where unstructured, time-varying aerodynamic parameters refer to stability and control derivatives, lift, drag and momentum coefficients, to say the least, which may adversely impact aircraft performance. Challenges related to problematic nonlinearities, coupling between lateral/longitudinal motion, uncertainty in aerodynamic parameters and in the system model, limit or prohibit applicability of known controller design techniques.

For these families of NextGen UAVs candidate navigation controller design methodologies should not be based solely on nominal plant and performance requirements—as has been typically the case so far for "traditional" UAV configurations—because the true plant may be partially unknown, or may belong to an admissible family of plants, or the plant may even change from one model to another from within the family of admissible plants. Any robust navigation controller design must ensure, for example, that closed-loop stability holds for any plant within the family of admissible plants, and that performance specifications are met throughout.

Therefore, this monograph aims to present a mathematically sound navigation controller design methodology with performance guarantees for systems with potentially unknown nominal plant models and with unstructured, time-varying uncertainties, centering on unconventional fixed-wing unmanned aircraft. The discussed foundational framework considers the different sources of (model, parameter, etc.) uncertainty, while the candidate controller is not designed solely based on nominal plant and performance requirements.

To the best of the authors' knowledge and experience, this monograph registers the first detailed methodology that lays the foundation for navigation controller design for a family of unconventional fixed-wing UAVs with aerodynamic uncertainty. The monograph will also provide the basis for comparison of other controller design approaches and their related applications.

The monograph is suitable for graduate students, scientists, and engineers who conduct research in the area of navigation control and controller design for nonlinear systems.

Denver, USA   Michail G. Michailidis
September 2019   Kimon P. Valavanis
                                                                       Matthew J. Rutherford

Acknowledgements This work was supported by National Science Foundation (NSF) Grant CMMI/DCSD 1728454.

Contents

Nomenclature

α	Angle of attack
β	Sideslip angle
χ	Course angle
δ_a	Aileron control surface
δ_e	Elevator control surface
δ_r	Rudder control surface
δ_t	Throttle
γ	Flight path angle
ℓ	Rolling moment
m	Pitching moment
n	Heading moment
ϕ	Roll angle of the UAV
ψ	Heading (yaw) angle of the UAV
ρ	Air density
θ	Pitch angle of the UAV
AR	Wing aspect ratio
b	Wingspan
c	Mean chord of the wing
C_ℓ	Rolling moment coefficient
C_m	Pitching moment coefficient
C_n	Heading moment coefficient
C_D	Drag coefficient
C_L	Lift coefficient
C_Y	Sideforce coefficient
d	Disturbance for control system diagram
e	Oswald efficiency factor
e	Tracking error for control system diagram
e_p	Performance tracking error for μ-synthesis design
F^b	UAV body frame
F^i	UAV inertial frame

F^s	UAV stability frame
F^{v_1}	UAV vehicle-1 frame
F^{v_2}	UAV vehicle-2 frame
F^v	UAV vehicle frame
F^w	UAV wind frame
F_T	Aircraft engine thrust
F_{i^b}	Total force along i^b axis
F_{j^b}	Total force along j^b axis
F_{k^b}	Total force along k^b axis
g	Gravity constant
h	Altitude of the aircraft
i^b	Axis of the body frame that points out the nose of the airframe
j^b	Axis of the body frame that points out the right wing of the airframe
k^b	Axis of the body frame that points out the belly of the airframe
m	Aircraft mass
p_d	Inertial down position of the UAV
p_e	Inertial east position of the UAV
p_n	Inertial north position of the UAV
S	Surface area of the wing
u	Velocity along i^b
v	Velocity along j^b
V_a	Airspeed
V_g	Ground Velocity
w	Velocity along k^b
p	Roll rate of the UAV
q	Pitch rate of the UAV
r	Heading rate of the UAV

Acronyms

ACL	Autonomous Control Level
ADS	Air Delivery System
ASU	Air Supply Unit
CC	Circulation Control
CIFER	Comprehensive Identification from Frequency Responses
CoG	Center of Gravity
GUI	Graphic User Interface
HJB	Hamilton Jacobi Bellman
HJPDI	Hamilton Jacobi Partial Differential Inequality
LFT	Linear Fractional Transformation
LPM	Linear Parametric Form
LPV	Linear Parameter Varying
LQG	Linear Quadratic Gaussian
LQR	Linear Quadratic Regulator
MIMO	Multi Input Multi Output
NED	North, East and Down
PD	Proportional Derivative
PID	Proportional Integral Derivative
RL	Reinforcement Learning
RPM	Revolutions Per Minute
SIL	Software In the Loop
SISO	Single Input Single Output
UAV	Unmanned Aerial Vehicle
UC^2AV	Unmanned Circulation Control Aerial Vehicle
UDP	User Datagram Protocol

Chapter 1
Introduction

Abstract This chapter intends to motivate the reader by presenting on overview of the investigated research problem. The rationale and motivation are discussed and a summary of observations and results is given. The challenge of robust control of a fixed-wing UAV with time-varying aerodynamic uncertainties is tackled by using and implementing μ-analysis and additive uncertainty weighting functions. The technique described herein can be generalized and applied to the class of new generation, unconventional UAVs, seeking to address uncertainty challenges regarding the aircraft's aerodynamic coefficients. A brief outline of the monograph is also given.

1.1 Motivation and Rationale

Almost twenty years after his loss against a cruel, heartless IBM supercomputer, former world chess champion and perhaps the greatest player in history, Garry Kasparov, delivered the message of not fearing intelligent machines [1]. The human versus machine controversy has been under discussion for decades, with different perspectives and opinions supporting both sides. One thing is certain though, humans triumph when machines triumph. With or without humans' consent, the future belongs to autonomous, intelligent and powerful machines which are gradually taking over, surpassing their own creators (Fig. 1.1).

In this sense, research and development in the field of design, autonomous navigation and control of unmanned aircraft has been rapidly growing over the last decade. The range of applications is vastly expanding, with aerial photography, traffic monitoring and military missions being some of the first to receive attention. As demand for Unmanned Aerial Vehicles (UAVs) increases, so does the need for intelligent and robust control systems that will guarantee a certain degree of autonomy.

UAVs are typically highly nonlinear underactuated systems; controller design presents challenges that need to be addressed and tackled. Challenges relate, among others, to problematic nonlinearities, coupling between lateral and longitudinal motions and uncertainty in aerodynamic parameters in the mathematical model (control and stability derivatives). When dealing with non-conventional UAV

© Springer Nature Switzerland AG 2020   1
M. G. Michailidis et al., *Nonlinear Control of Fixed-Wing UAVs*
with Time-Varying and Unstructured Uncertainties, Springer Tracts
in Autonomous Systems 1, https://doi.org/10.1007/978-3-030-40716-2_1

Fig. 1.1 Garry Kasparov's TED talk "Don't Fear Intelligent Machines" [1]

designs, such inherent uncertainties either limit or prohibit applicability of known controller design techniques. Viewed from this perspective, controller design for non-conventional UAVs (i.e., new generation UAVs) requires consideration of unstructured parameters, model uncertainty and an advanced controller design framework.

As mentioned in [2, 3], modeling the aircraft aerodynamic coefficients raises the fundamental question of what the mathematical structure of the model should be. Although a complicated model structure can be justified for accurate description of the aerodynamic forces and moments, it is not always clear what the relationship between model complexity and information in the measured data should be. If too many model parameters are sought for a limited amount of data, reduced accuracy of estimated parameters is expected, or the attempts to estimate all the parameters in the model might fail. Aircraft system identification is a complex process and the final values for the estimated aerodynamic parameters are usually within some certain error bounds. Therefore, even in the classical, conventional UAV case, aerodynamic/model uncertainty should be taken into consideration for flight control and navigation purposes.

Regardless of the nature of the system to be controlled, a candidate controller cannot be designed solely on the basis of nominal plant and performance requirements. The true plant is (partially) unknown and it must belong to an admissible family of plants, as Fig. 1.2 shows. Model uncertainty must be addressed and tackled. Robust controller design ensures that closed-loop stability holds for any plant within this family, and that performance specifications are met. In real-life problems, a nominal model is an intentional approximation to reality. However, if model uncertainty is not accounted for and if the nominal plant model is exclusively used, the nominal feedback design might not be stable and only strict performance specifications will be met.

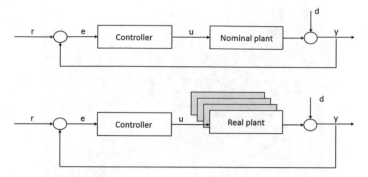

Fig. 1.2   Nominal versus real plant control system diagram

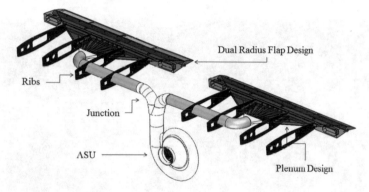

Fig. 1.3   CAD design of CC system [4]

This research is motivated by the challenge to design, model, build, control and test a small-scale Unmanned Circulation Control Aerial Vehicle (UC^2AV), which is the first of its kind with proven flight capability [4]. Circulation Control (CC) is an active flow control technique that is proven to be an efficient method for lift augmentation resulting in improved aerodynamic efficiency, runway reduction during takeoff/landing, smoother landing, enhanced payload capabilities and delayed stall.

For experimentation, validation and verification, a stock RMRC Anaconda has been integrated with a CC system on-board (Figs. 1.3 and 1.4), which operates on demand according to the ongoing mission. The UC^2AV has been designed to perform missions with different flight requirements and mission adaptation gives the ability to the end user to operate a single UAV for multiple applications.

Operating the CC system on demand results in direct changes of the aircraft control and stability derivatives during flight. Preliminary research has shown a reduction in take-off distance by 54% compared to the conventional UAV as depicted in Fig. 1.5.

Fig. 1.4 The UC^2AV [4]

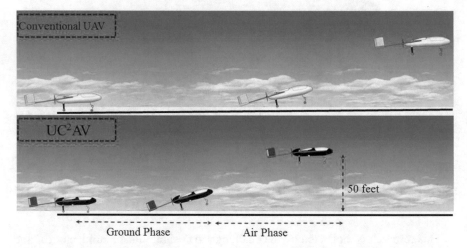

Fig. 1.5 Take-off performance behavior of the UC^2AV compared to a conventional UAV [4]

1.2 Problem Statement

The overall controller design problem statement is summarized in Fig. 1.6. The objective is to build a navigational and stability controller that will regulate the attitude angles and the position of the platform, achieving trajectory tracking while taking into consideration uncertain, on-demand alteration of the vehicle's aerodynamic parameters. The variables to be controlled are divided into lateral and longitudinal motion as the standard convention dictates. However, the proposed controller design framework will not be based on the motion decoupling assumption that is followed for conventional (commercial) fixed-wing airplanes.

The feature of uncertain, time-varying aerodynamic characteristics on the aircraft equations of motion is what separates this study from existing ones in the field of autonomous flight controller design. Therefore as a first step, this feature will be

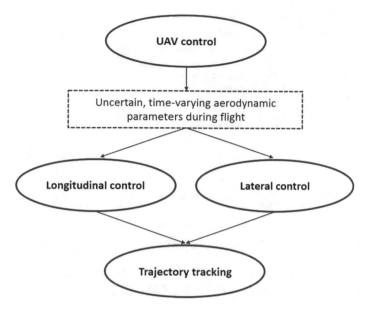

Fig. 1.6 UAV control strategy

clearly explained and justified. Based on this, a valid and realistic UAV model will be proposed and investigated, leaving the actual controller design process as the final stage.

1.3   Method of Approach

The proposed solution to the stated UAV controller design problem is a hybrid control architecture consisting of an inner/outer loop structure which can be seen in Fig. 1.7. The inner loop will employ a dynamic inversion controller for partial linearization of the UAV dynamics, whereas the outer loop will be a μ-synthesis controller to ensure robustness against external disturbances and an on-demand alteration of the UAV aerodynamic parameters [5, 6]. The effect of uncertainty in aerodynamic character-istics of the platform will be tackled by the use of additive uncertainty weighting functions which will be used in the μ-synthesis and analysis design process. Lastly, sensor noise will be considered and complementary filter design will be incorporated in the controller framework for accurate estimation of the state variables that will require measurement, with an ultimate goal of minimizing the tracking error e_p.

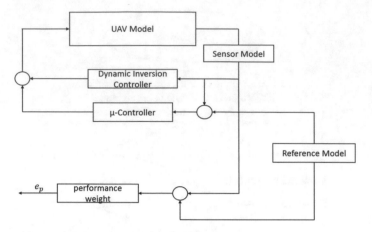

Fig. 1.7 UAV proposed controller design framework

1.4 Summary of Observations and Results

Unlike existing work in the field of navigation and control of UAVs, this monograph is specific to controller design for non-conventional, new generation aircraft subject to aerodynamic uncertainty, with time-varying aerodynamic parameters during flight. To the best of the author's knowledge, this is the first technical, comprehensive study, laying the foundation for autonomous navigation of UAVs in the presence of aerodynamic uncertainty, providing a basis for comparison for all the control design approaches and their related applications.

Although published material tackles important aspects of navigation and control as well as UAV applications, there is no insight on controller design techniques for UAVs with time-varying aerodynamic parameters and related challenges. The major contributions of this monograph are the following:

- Establish and highlight the need for a new UAV controller design framework.
- Derive a valid and realistic theoretical mathematical model for the aircraft's aerodynamic coefficients.
- Design a robust, nonlinear controller capable of handling time-varying aerodynamic parameters during flight.
- Integrate the controller framework with sensor noise and sensor models utilizing complementary filters.
- Validate the control scheme using MATLAB/Simulink and the X-Plane flight simulator.

1.5 Outline of the Monograph

The remainder of this monograph is organized as follows: Chap. 2 presents a literature review, investigating the controller design techniques found in the literature and evaluating their distinctive advantages and setbacks. Chapter 3 gives necessary background related to UAV equations of motion and highlights the source of the research problem, establishing the relation between aerodynamic changes and the aircraft control and stability derivatives. Chapter 3 also presents motivating existing or potential real-life applications where this work may prove useful for researchers. Chapter 4 covers the controller design algorithm with a step-by-step, detailed analysis and inspection. Validation results can be found in Chap. 5, where the proposed controller design framework is employed for robust control of the UC^2AV. Finally, concluding remarks, challenges and future work are included in Chap. 6.

References

1. Kasparov G (2017) Don't fear intelligent machines: work with them. ICGA J 39(2):126–128
2. Klein V, Morelli EA (2006) Aircraft system identification: theory and practice. American Institute of Aeronautics and Astronautics, Reston
3. Tischler MB, Remple RK (2006) Aircraft and rotorcraft system identification. AIAA education series. American Institute of Aeronautics and Astronautics, Reston
4. Kanistras K, Rutherford MJ, Valavanis KP (2018) Foundations of circulation control based small-scale unmanned aircraft. Springer, Berlin
5. Michailidis MG, Agha M, Kanistras K, Rutherford MJ, Valavanis KP (2017) A controller design framework for a NextGen circulation control based UAV. In: IEEE conference on control technology and applications (CCTA), pp 1542–1549
6. Michailidis MG, Kanistras K, Agha M, Rutherford MJ, Valavanis KP (2017) Robust nonlinear control of the longitudinal flight dynamics of a circulation control fixed wing UAV. In: IEEE conference on decision and control (CDC), pp 3920–3927

Chapter 2
Literature Review

Abstract This chapter provides a technical overview and the necessary background for existing controller synthesis methods that have been applied for navigation and control of UAVs. These include linear controllers (PID, LQR, LQG, etc.), backstepping, sliding mode, nonlinear model predictive, adaptive, dynamic inversion, fuzzy logic and neural networks, gain scheduling, H_∞ and μ-synthesis [1, 2]. The distinctive advantages and drawbacks for each technique are investigated with respect to applicability to the family of new generation UAVs with time-varying aerodynamic characteristics.

2.1 Published Surveys Review

Eleven surveys have been published to date, exploring research in the areas of autopilot hardware and software, control techniques, motion planning, collision avoidance, traffic surveillance, imagery collection, communication networks and vision-based navigation. This section presents a summary of contributions of existing surveys.

Published in 2004, "*Control and Perception Techniques for Aerial Robotics*" [3], is mostly focused on perception techniques, reviewing methods that have been applied to aerial robotics including different vehicle platforms and flight control hardware. It provides a brief survey of control architectures and computer vision techniques. It covers a broad range of UAVs, but little emphasis is placed on controller design methodologies.

Published in 2005, "*A Survey of Unmanned Aerial Vehicles (UAV) for Traffic Surveillance*" [4], presents a survey of research activities in several universities around the world in the area of application of UAVs in traffic surveillance. A summary of research projects, vehicle platforms and research objectives is provided with respect to traffic sensing and management.

Published in 2009, "*A Survey of Autonomous Control for UAV*" [5], surveys the autonomous control concept and Autonomous Control Level (ACL) metrics that can measure autonomy of UAVs. The constraint conditions and realizations of the three basic levels of UAV system autonomy (execution, coordination and organization)

© Springer Nature Switzerland AG 2020

M. G. Michailidis et al., *Nonlinear Control of Fixed-Wing UAVs*
with Time-Varying and Unstructured Uncertainties, Springer Tracts
in Autonomous Systems 1, https://doi.org/10.1007/978-3-030-40716-2_2

are studied comprehensively. The key hardware and software technologies for multi-tasking are modularized depending on mission requirements.

Published in 2009, "*A Survey of Collision Avoidance Approaches for Unmanned Aerial Vehicles*" [6], focuses on collision avoidance approaches deployed for unmanned aerial vehicles. The collision avoidance concept is introduced together with proposing generic functions carried by collision avoidance systems. The design factors of the sense and avoid system are explained in detail and based on these, several typical approaches are categorized.

Published in 2010, "*A Survey of Motion Planning Algorithms from the Perspective of Autonomous UAV Guidance*" [7], provides an overview of existing motion planning algorithms while adding perspectives and practical examples from UAV guidance approaches. It emphasizes practical methods and provides a general perspective on the particular problems arising with UAVs.

Published in 2010 in the IJCAS journal, "*Autopilots for Small Unmanned Aerial Vehicles: A Survey*" [8], contains a survey of autopilot systems intended for use with small or micro UAVs. Several typical commercial off-the-shelf autopilot packages are compared in detail and some research autopilot systems are introduced. Concluding remarks are made with a summary of the autopilot market and a discussion on the future directions.

Published in 2011, "*A Survey of Unmanned Aerial Vehicle (UAV) Usage for Imagery Collection in Disaster Research and Management*" [9], provides a review of utilization of UAVs for imagery collection for disaster monitoring and management. A review of papers regarding data acquisition and assessment prior, during and after disaster events is presented.

Published in 2012, "*Survey of Motion Planning Literature in the Presence of Uncertainty: Considerations for UAV Guidance*" [10], surveys motion planning algorithms that can be applied on UAVs and that can deal with the primary sources of uncertainty arising in real world missions. Emphasis is placed on uncertainties in vehicle dynamics and environment knowledge, investigating optimal, model predictive and Lyapunov techniques for the first as well as A^* and D^* planning techniques for the second.

Published in 2014, "*A Survey of Small-Scale Unmanned Aerial Vehicles: Recent Advances and Future Development Trends*" [11], provides a detailed overview of advances of small-scale UAVs including platforms and scientific research areas. The evolution of the key elements, including on-board processing units, navigation sensors, mission-oriented sensors, communication modules, and ground control station is presented and analyzed. Finally, the future of small-scale UAV research, civil and military applications are forecasted.

Published in 2016, "*Survey of Important Issues in UAV Communication Networks*" [12], focuses on the issues of routing, seamless handover and energy efficiency in UAV networks. A categorization of UAV networks and an examination of important characteristics like topology, control, and client server behavior is carried out. Requirements from the routing protocols unique to UAV networks and the need for disruption tolerant networking are also discussed.

Published in 2018, *"A survey on vision-based UAV navigation"* [13], presents a comprehensive literature review of the vision-based methods for UAV navigation. Specifically, it focuses on visual localization and mapping, obstacle avoidance and path planning, which compose the essential parts of visual navigation. Furthermore, an insight into the prospect of UAV navigation and the challenges to be faced is given.

There is no existing technical and detailed study, evaluating the control techniques for navigation and control of the family of new generation aircraft. This study aims to serve as a state-of-the-art and establish the foundational methodology to design controllers for complex, uncertain UAV systems with a particular focus on the significant challenge of aerodynamic uncertainty.

2.2 Linear Control

PID controllers are a type of single-input/single-output (SISO) control structure. A great advantage of PID controllers is that they can be easily implemented and they require low computational effort on-board the UAV [8]. It is also relatively easy to build on top of PIDs, in cascaded loops as in [14], meaning that they can be effectively combined with other synthesis methods. On the other hand, as stated in [8], PID techniques are non-model based and they lack robustness. Their non-model based characteristic can be considered as an advantage, but in the case of a UAV with time-varying aerodynamic uncertainties, tuning the PID gains can become a rather difficult task due to model uncertainty.

Linear Quadratic Gaussian (LQG) and Linear Quadratic Regulator (LQR) are optimal feedback controllers based on minimizing predefined cost functions and can be used both for SISO and MIMO (multi-input/multi-output) structures. LQG control can also operate in the presence of white noise.

These techniques can be used for multi-variable systems but due to their iterative nature, the control input vector may be hard to determine [15]. Additionally, input constraints of the system are not taken into consideration. An application of LQR for UAV flight control can be seen in [16], presenting a 3D LQR based landing controller that accurately lands the vehicle on a runway.

Every linear technique is based on the fact that the studied system model is linear. This means that even if the actual system behaves in a nonlinear way, in order to apply linear methods, one has to linearize the given model around some specific operating condition. Linearization can be convenient but it has local validity, only in a certain neighborhood around the specified condition.

State of the art in linear controller design for fixed-wing UAV tackles the challenges of PID auto-tuning [17] and model uncertainty and robustness by using gain-scheduling [18]. Studies comparing PID, LQR, adaptive, neural, fuzzy and backstepping designs can be found in [19, 20]. Adaptive neuro-fuzzy techniques are proven to be more efficient, indicating that linear controllers cannot provide robust performance guarantees in presence of large-scale aerodynamic uncertainties.

2.3 Backstepping

Backstepping has been widely used for UAV control due to its recursive nature; its foundation lies in Lyapunov analysis [21]. One requirement for backstepping to be applied is the system to be put in strict feedback form [22], see (2.1). Virtual control inputs are generated in order to account for the deficit between the number of system states and the number of actual control inputs. The design can benefit from useful nonlinearities by appropriately choosing these virtual control inputs.

$$\dot{x} = f(x) + g(x)\xi_1$$
$$\dot{\xi}_1 = f_1(x, \xi_1, \xi_2)$$
$$\dot{\xi}_2 = f_2(x, \xi_1, \xi_2, \xi_3)$$
$$\vdots$$
$$\dot{\xi}_{k-1} = f_{k-1}(x, \xi_1, \ldots, \xi_k)$$
$$\dot{\xi}_k = f_k(x, \xi_1, \ldots, \xi_k, u)$$

(2.1)

The general concept of backstepping can be seen in Fig. 2.1, for the simplest system $\dot{z} = f(z) + g(z)\xi$, $\dot{\xi} = u$. The asymptotically stabilizing control law $\phi(z)$ is "backstepped" through the integrator. The primary challenge for backstepping control designs is finding a potential Lyapunov candidate function.

Putting the UAV equations of motion into a strict feedback form as in (2.1) requires a set of a-priori assumptions related to the aircraft aerodynamics [23]. As far as new generation UAVs are concerned, this is acceptable but not preferable. Furthermore, backstepping is a robust technique but it is sensitive to aerodynamic parameter variation. Researchers have employed more sophisticated control architectures such as adaptive, for trajectory tracking [24] and disturbance rejection [25, 26], or incremental (sensor-based) backstepping [27] to robustify the technique and make it more versatile. An interesting comparison of backstepping, PID and fuzzy PID can be found in [28] for UAV path planning, concluding that fuzzy PID provides superior performance.

2.4 Sliding Mode

Sliding mode is a nonlinear control method designed to constrain the system states to a certain manifold or sliding surface. In its ideal setup, sliding mode requires the control input to oscillate with very high frequency but this may not be achievable for every dynamic system [29]. The trajectory of the system states does not always stay on the sliding surface but instead, it may oscillate around the surface due to delays in control switching in what is called chattering [30, 31]. Sliding mode generates

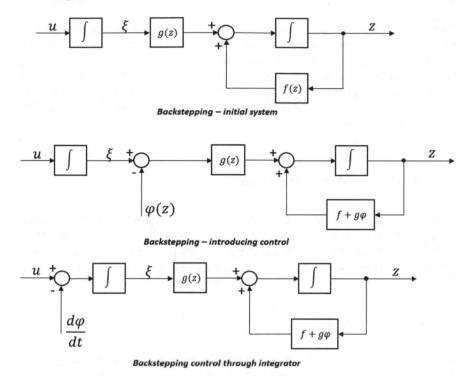

Fig. 2.1 The backstepping concept

discontinuous control laws, raising questions about the existence and uniqueness of solutions and the validity of Lyapunov analysis.

The mathematical objective of sliding mode control is to transform a system of the form $\dot{x} = f(x) + B(x)(G(x)u + \delta(t, x, u))$ into a system in a regular form as in (2.2) by utilizing an appropriate change of variables.

$$
\begin{aligned}
\dot{\eta} &= f_\alpha(\eta, \xi) \\
\dot{\xi} &= f_b(\eta, \xi) + G(x)u + \delta(t, x, u)
\end{aligned}
\tag{2.2}
$$

Parameter x is the state vector, u is the control input vector, f and B are sufficiently smooth functions and G, δ are uncertain functions. The sliding manifold $s = \xi - \phi(\eta) = 0$ is then designed so that when the motion is restricted to the manifold, the reduced-order model $\dot{\eta} = f_\alpha(\phi(\eta))$ has an asymptotically stable equilibrium point at the origin. This is achievable for attitude control of a new generation UAV because sliding mode guarantees robustness against aerodynamic/model uncertainty with a given upper bound.

Applications of adaptive sliding mode control for fixed-wing UAVs can be found in [32, 33], where adaptation is employed to deal with the effect of chattering and optimize robustness against model uncertainty. In [34], an adaptive PD controller is designed with the adjustment mechanism following the gradient-based MIT rule. Recent advances in the field of continuous sliding mode control of UAVs are established in [35–38], proposing a technique that eliminates the effect of chattering. Finally, a study comparing backstepping, sliding mode and backstepping with sliding mode control can be found in [39], concluding that backstepping with high order sliding mode achieves superior performance with a better minimization of the chattering effect.

2.5   Nonlinear Model Predictive

Nonlinear model predictive control is a technique that can predict the future behavior of the system and allows for on-line implementation. It is based on the concept of repetitively solving an optimization problem involving a finite time horizon and a dynamic mathematical model [40]. The goal is to minimize a cost function of the form

$$J[u(t), x(t)] = \int_0^T l(x(t), u(t), t)dt + S(x(T), T) \tag{2.3}$$

where T is the time horizon, function l denotes the stage cost and function S represents the terminal cost, subject to the physical constraints

$$u_{min} \le u(t) \le u_{max} \quad , \quad g(x(t), u(t), t) \le 0 \tag{2.4}$$

with the dynamic mathematical model described by the ordinary differential equation $\frac{d}{dt}x(t) = f(x(t), u(t), t)$. Solving this differential equation for a new generation UAV, either analytically or numerically, is a challenging task. The UAV control and stability derivatives (function f) will be uncertain and time-varying, so the process will be computationally intensive for on-board implementation. Nonlinear optimization of the cost function (2.3) requires accurate sensor measurement of the state vector $x(t)$, or alternatively, employment of linear model predictive control approaches [41, 42].

However, the feature that prohibits applicability of nonlinear model predictive on a new generation UAV is dependence on system knowledge. In principle, nonlinear model predictive designs cannot handle large scale, time-varying uncertainties because system knowledge is required for model prediction. A low-level kinematic model of the UAV dynamics is utilized in [43] to design a high-level controller for path following. An adaptive nonlinear model predictive approach that varies the conventional fixed horizon according to the path curvature profile is proposed in [44].

A large body of literature, including recent advances such as [45–47], utilizes a UAV kinematic model to achieve trajectory tracking with a nonlinear model predictive design due to its ability to explicitly handle the control input and system state constraints highlighted in (2.4). Although this approach is generally applicable

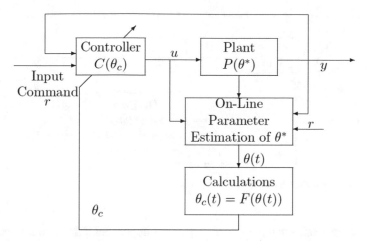

Fig. 2.2 Indirect adaptive control [54]

for a new generation UAV, the controller will be non-model based, meaning that the time-varying control and stability derivatives will not be taken into consideration. An advanced architecture that guarantees stability properties in presence of time-varying uncertainties would be more suitable.

2.6 Adaptive

The design of a controller that can alter or modify the behavior and response of an unknown plant to meet certain performance requirements can be a tedious and challenging problem in many control applications. By definition, to adapt means to change (oneself) so that one's behavior will conform to new or changed circumstances. Adaptive control seeks to address issues of parametric or environmental uncertainties based on the Lyapunov concept of stability [48, 49].

Unknown parameter vectors are defined and estimated so that Lyapunov stability is guaranteed, following two main approaches, the indirect (Fig. 2.2) and the direct adaptive control (Fig. 2.3). Adaptive control enables a wide operation range during flight as demonstrated in [50–53], where adaptive is used to robustify backstepping, neural and fuzzy designs against model uncertainty and unmodeled dynamics.

Adaptive control strategies can be categorized according to whether the controller parameters are tuned continuously in time or switched between discrete values at specified instants. The first category refers to the classical, deterministic adaptive control and has some inherent limitations due to dependence on an identified plant model. This issue becomes severe if robustness and high performance is sought. In the second case, switching can be performed among controllers of different structures, resulting in a design that is independent of plant identification accuracy and other prior assumptions [55].

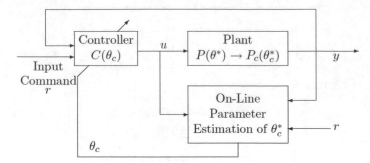

Fig. 2.3 Direct adaptive control [54]

A major setback for the applicability of traditional adaptive control for a new generation UAV is limited flexibility of the unknown parameter vector for robust controller design purposes.

Switching multi-model adaptive control provides a more robust alternative compared to the classical adaptive control approach. The idea lies behind switching between stabilizing and destabilizing controllers from a predefined set to achieve asymptotic stability. Switching among candidate controllers is orchestrated by a high-level decision maker called a supervisor. The supervisor updates controller parameters when a new estimate of the process parameters becomes available, similarly to the adaptive control paradigm, but these events occur at discrete instants of time [56, 57]. This results in a hybrid closed-loop system. The general view of a switching adaptive control system, in which the control action is based on the learned characteristics of the process (plant) is depicted in Fig. 2.4.

If the parametric uncertainty is described by a continuum, one has the choice of working with a continuous or a discrete family of controllers. In this case, one needs to ensure that every admissible process model is satisfactorily controlled by at least one of these controllers. The switching algorithms that seem to be the most promising are those that evaluate the potential performance of each candidate controller on-line and use this to direct their search. Comprehensive examples of fuzzy adaptive control for switched systems can be found in [58–60].

The mathematical foundation and the ground for the design of switched adaptive control systems has been well established in numerous works over the last two decades. A recent application for robotic manipulators can be seen in [61]. Nevertheless, real-life aerospace applications of the switching adaptive control strategy are yet to be seen. The supervisory control system framework requires thorough analysis and understanding, not to mention the potential computational burden the control systems engineer might have to face for a real-time application.

One last limitation of this approach is the speed of switching between candidate controllers, occurring based on observed system data. For instance, designing a switched adaptive controller for a fighter aircraft, or a morphing aircraft with on-demand configuration, might prove to be a significant challenge.

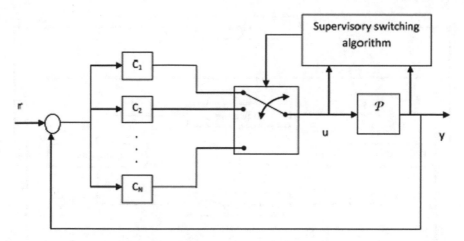

Fig. 2.4 Adaptive control architecture consisting of a switching controller and a supervisory controller block [56]

2.7 Dynamic Inversion

Dynamic inversion or feedback linearization is a method seeking to transform the nonlinear system dynamics into an equivalent, fully or partially linear form through some algebraic transformation. Given a system of the form $\dot{x} = f(x) + g(x)u$, if the control law $u = g^{-1}(x)[-f(x) + ax]$ is applied for some constant a, the initial nonlinear system can transform into a linear one. This simple idea summarizes the concept behind dynamic inversion. Linear transformation can be achieved by somehow inverting the nonlinear UAV dynamics and solving the puzzle of motion decoupling [62]. By applying dynamic inversion, one controller is capable of handling the entire flight regime.

Recent applications of dynamic inversion for unmanned aircraft systems can be found in [63–66], where observer-based dynamic inversion is used to account for input constraints and inaccurate sensor measurements. Dynamic inversion can be used in cascaded designs for performance tuning. For instance, dynamic inversion is robustified by the use of gain scheduling in [67] after linearizing the system to handle the complex UAV system dynamics. Additionally, dynamic inversion can efficiently serve as an inner-loop control law for H_∞ and μ-synthesis designs that will be analyzed in a subsequent section. However, the control law u is implementable only if the system is precisely known, which is a significant limitation for application on a new generation UAV. This would require accurate measurement of the UAV attitude angles, linear velocities and angular rates, as well as a precise feedback of the time-varying control and stability derivatives during flight.

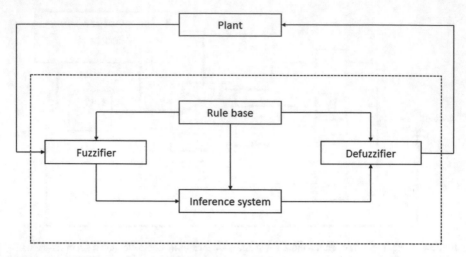

Fig. 2.5 Architecture of fuzzy logic control system

2.8   Fuzzy Logic/Neural Networks

Fuzzy logic control is a model-free, knowledge based technique which tries to mimic the way humans think and make decisions by creating a set of rules that are used by the controller to analyze the input and to determine the appropriate output. The basic concept of a fuzzy control system is depicted in Fig. 2.5 and the main steps for a fuzzy logic control algorithm are given below.

   (i) Define the linguistic variables and terms (initialization)
  (ii) Construct the membership functions (initialization)
 (iii) Construct the rule base (initialization)
 (iv) Convert crisp input data to fuzzy values using the membership functions (fuzzi-fication)
  (v) Evaluate the rules in the rule base (inference)
 (vi) Combine the results of each rule (inference)
(vii) Convert the output data to non-fuzzy values (defuzzification).

Membership functions in fuzzy logic control systems are used in the fuzzification and defuzzification process to convert non-fuzzy input values to linguistic terms and vice versa. Fuzzy logic is a model-free, intuitive design that can be built up and trained for specific applications. The fuzzy logic UAV controller follows a (if event A, then event B) framework based on the rule base, meaning that it indirectly deals with aerodynamic uncertainties in the UAV model [68, 69]. In the case of new generation aircraft however, where aerodynamic uncertainties have time-varying structure, several simulation or flight tests will be needed to train the system and the designed controller to achieve robust performance for every on-demand change of the aerodynamic coefficients (event A). Consistency of rules and system tuning parameters (inference, fuzzification and defuzzification) have to be investigated because system stability and optimization can only occur experimentally.

Unlike fuzzy, neural networks are a learning based method that seeks to mimic the human central nervous system by utilizing input-output data to program the neurons in a network. A three-layer neural network structure to account for aerodynamic uncertainties in the UAV model can be found in [70]. Recent genetic neuro-fuzzy applications on fixed-wing aircraft are reported in [71, 72] to deal with lack of modeling and flight uncertainties. State of the art in intelligent (fuzzy and neural networks) flight control systems for small aerial vehicles is discussed in [73]. The challenges of computational demand, online learning and uncertainty in data representation are highlighted for the still growing field of intelligent aerial robotics.

2.9 Gain Scheduling

Gain scheduling is a switching strategy between a finite number of linear controllers each corresponding to a linear model of the aircraft dynamics near a design trim condition. The idea behind designing a gain scheduled controller for a nonlinear plant, illustrated in Fig. 2.6 and taken from [74], can be described as a four step procedure as follows:

(i) The first step is to compute a linear-parameter-varying (LPV) model for the aircraft. The traditional approach in this area is based on Jacobian linearization of the nonlinear plant about a family of equilibrium points, also called operating points or set points. This yields a parametrized family of linearized plants and forms the basis for linearization scheduling. A detailed comparative study can be found in [75], where LPV models for the Boeing 747–100/200 are derived and evaluated.

(ii) The second step is to use linear design methods to design linear controllers for the linear parameter-varying plant model that arises. This design process results in a family of linear controllers corresponding to the linear-parameter-dependent plant. Traditionally, the designs are such that for each fixed value of the parameter, the linear closed-loop system exhibits desirable performance.

(iii) The third step includes the actual gain scheduling. A family of linear controllers is implemented so that the controller coefficients (gains) are varied (scheduled) according to the current value of the scheduling variables.

(iv) Performance assessment is the final step. Desired performance guarantees might be part of the design process but typically, the local stability and performance properties of the gain scheduled controller might be subject to analytical investigation, while the nonlocal performance evaluation might require simulation studies.

Gain scheduling employs powerful linear design tools on difficult nonlinear problems. Gain scheduled controllers preserve well-understood linear intuition, in con-

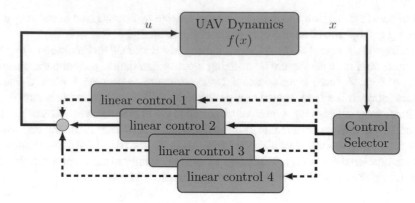

Fig. 2.6 Functionality of gain scheduling [76]

trast to nonlinear control approaches that involve coordinate transformations. Moreover, gain scheduling enables the controlled system to respond rapidly to changing operating conditions. Last but not least, the computational burden of linearization scheduling approaches is often much less than other nonlinear design approaches. Applications of gain-scheduling for morphing aircraft can be found in [77, 78] whereas a detailed gain-scheduled flight control design is performed in [79].

Limitations of gain scheduling for control of a new generation UAV include the large number of flight conditions that need to be considered and also the need for the transition between the models to be smooth. Stability can be assured only locally and in a slow-variation setting and usually there are no performance guarantees. This presents a bottleneck in the case of a UAV with rapidly changing aerodynamic parameters.

2.10 H_∞ and μ-Synthesis

Linear H_∞ is a type of multi-variable, robust, model-based control and its major advantage over linear techniques is its robustness in presence of model uncertainties. Given a linear, time-invariant system Σ as depicted in Fig. 2.7, with w being the exogenous input, z being the corresponding output and u, y representing regular inputs and outputs, a control law $u = F_1 x + F_2 w$ is sought that will minimize the H_∞ norm of the overall transfer matrix over parametric uncertainties Σ_K [80, 81].

Nonlinear H_∞ is based on the same optimization concept and it is transformed into a nonlinear technique through the use of a dynamic inversion inner-loop control law for linearization of the dynamics [82–84]. The Hamilton–Jacobi partial differential inequality (HJPDI) that can be found in [85, 86] is another alternative.

In a nutshell, given a nonlinear system $\dot{x} = f(x) + g_1(x)d + g_2(x)u$, the HJPDI approach attempts to solve the differential inequality shown in (2.5).

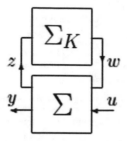

Fig. 2.7 H_∞ control system [81]

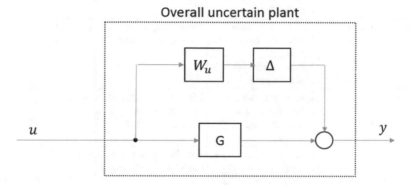

Fig. 2.8 Nominal plant with additive uncertainty [91]

$$\left(\frac{\partial E}{\partial x}\right)^T f + \frac{1}{2}\left(\frac{\partial E}{\partial x}\right)^T \left(\frac{1}{\gamma_1^2}g_1 g_1^T - g_2 W_E^{-2} g_2^T\right)\left(\frac{\partial E}{\partial x}\right) + \frac{1}{2}h_1^T h_1 < 0 \quad (2.5)$$

for some positive C^1 function E, output signal h_1 and weighting function W_E, and then make use of the nonlinear bounded real lemma. The μ-synthesis method is an extension to the H_∞ design because it is a doubly-iterative optimization process with respect to:

1. the H_∞ compensator $K(s)$
2. the $D(s)$ scales.

An optimal H_∞ compensator $K(s)$ is designed and the scales $D(s)$ are optimized so that the robust complex-μ test shown in (2.6) is satisfied, for the system's overall transfer matrix M [87].

$$\mu(M) = \mu(DMD^{-1}) \leq \sigma_{max}(DMD^{-1}) \leq 1, \ \forall \omega \in \mathbb{R} \quad (2.6)$$

The μ-synthesis framework allows model uncertainties and system perturbations to enter the design in a multiplicative or additive fashion [88]. The authors have imple-

Table 2.1 Summary of non-qualifying control techniques

Method	Advantages	Disadvantages	References
Linear control	Straightforward design, low computational effort, cascaded loops	Local validity, robustness issues	[14–20]
Backstepping	Efficient for underactuated systems	Strict feedback form, sensitive to parameter variation	[21–28]
Sliding mode	Sliding manifold, robust against model uncertainty	Discontinuous control law, effect of chattering	[29–34, 39]
Nonlinear model predictive	Can predict future behavior of the system, can handlesystem and input constraints	Dependent on the system knowledge	[40–47]
Adaptive	Can handle unknown parameters, wide operation range available	Limited flexibility of the unknown parameter vector	[48–54]
Dynamic inversion	One control law needed, no motion decoupling required	Highly dependent on the system knowledge	[62–67]

mented a novel, nominal plant with additive uncertainty (Fig. 2.8), where an uncertainty range such as $C_{L_1} \leq C_L \leq C_{L_2}$ is utilized to compensate for time-varying, lift coefficient, aerodynamic uncertainties of a new generation aircraft. The approach was based on research reported in [89, 90], where a μ-synthesis controller is designed for a 4-wheel vehicle and then extended for application on fixed-wing aircraft.

Both H_∞ and μ-synthesis can deal with nonlinear, multi-variable systems and can also handle UAV time-varying aerodynamic uncertainties through an off-line definition of the uncertainty interval. Performance specifications, disturbances in several locations in the feedback loop and actuator models are also considered. The entire μ-synthesis controller design can be simplified, validated and supported by existing MATLAB software such as [87, 88]. A possible increase of the complexity is anticipated as the model dimension increases and the controller will only be optimal with respect to a predefined cost function and not to other common measures such as settling time.

Table 2.2 Summary of proposed controller design frameworks

Method	Advantages	Challenges	References
Continuous sliding mode	Robust against model uncertainty, minimization of chattering	Smoothness of control inputs	[35–38]
Switching adaptive control	Learning-based supervisory control algorithm, direct performance evaluation of candidate controllers, asymptotic stability guaranteed	Potential computational burden, no existing real-lifeapplications, speed of switching	[55–61]
Fuzzy logic, neural networks	Model free method, intuitive design, can be built up and trained	Intensive simulation or flight tests needed to be trained	[68–73]
Gain scheduling	Simplification of controller design, rapid response to changing parameters, computationally efficient	Large number of flight conditions need to be considered, smooth transition	[74–79]
H_∞ and μ-Synthesis	Robust in presence of uncertainties, performance specs and actuator models are considered	Increase of complexity as dimension increases, controller is optimal with respect to a predefined cost function	[80–91]

2.11 Theoretical Comparison of Controller Designs

This section gives a comprehensive summary of the literature review performed in this chapter, providing a basis for comparison for researchers that is divided into two concise tables. Table 2.1 gives a general overview of the non-qualifying techniques, containing the advantages and disadvantages for each method that is addressed and evaluated. Table 2.2 summarizes the promising and applicable control architectures for new generation UAVs, highlighting potential challenges as well as the respective references used to justify our claim.

2.12 Remarks

The literature review that has been performed for all existing controller design techniques supports the need for new and more versatile methods that are able to handle time-varying, model-based uncertainties. Old-fashioned techniques are fading and learning-based, switching and universally robust methods are evolving and slowly but steadily replacing their predecessors. Specifically, given the limited published material that revolves around μ-synthesis and robust control of time-varying uncertainties, the need to introduce a rigid, mathematically solid μ-synthesis framework is the most significant highlight of this chapter. With that in mind, UAV modeling fundamentals are covered in the next chapter.

References

1. Michailidis MG, Rutherford MJ, Valavanis KP (2019) A survey of controller designs for new generation UAVs: The challenge of uncertain aerodynamic parameters. Int J Control, Autom Syst
2. Michailidis MG (2019) Nonlinear controller design for UAVs with time-varying aerodynamic uncertainties. Doctoral dissertation, University of Denver
3. Ollero A, Merino L (2004) Control and perception techniques for aerial robotics. Annu Rev Control 28(2):167–178
4. Puri A (2005) A survey of unmanned aerial vehicles (UAV) for traffic surveillance. University of South Florida, Department of computer science and engineering
5. Chen H, Wang XM, Li Y (2009) A survey of autonomous control for UAV. In: IEEE international conference on artificial intelligence and computational intelligence (AICI), pp 267–271
6. Albaker BM, Rahim NA (2009) A survey of collision avoidance approaches for unmanned aerial vehicles. In: International conference for technical postgraduates (TECHPOS), pp 1–7
7. Goerzen C, Kong Z, Mettler B (2010) A survey of motion planning algorithms from the perspective of autonomous UAV guidance. J Intell Robot Syst 57(1–4):65
8. Chao H, Cao Y, Chen Y (2010) Autopilots for small unmanned aerial vehicles: a survey. Int J Control, Autom Syst 8(1):36–44
9. Adams SM, Friedland CJ (2011) A survey of unmanned aerial vehicle (UAV) usage for imagery collection in disaster research and management. In: International workshop on remote sensing for disaster response
10. Dadkhah N, Mettler B (2012) Survey of motion planning literature in the presence of uncertainty: considerations for UAV guidance. J Intell Robot Syst 65(1–4):233–246
11. Cai G, Dias J, Seneviratne L (2014) A survey of small-scale unmanned aerial vehicles: recent advances and future development trends. Unmanned Syst 2(2):175–199
12. Gupta L, Jain R, Vaszkun G (2016) Survey of important issues in UAV communication networks. IEEE Commun Surv Tutor 18(2):1123–1152
13. Lu Y, Xue Z, Xia GS, Zhang L (2018) A survey on vision-based UAV navigation. Geo-Spat Inf Sci
14. Adami TM, Zhu JJ (2011) 6DOF flight control of fixed-wing aircraft by trajectory linearization. In: IEEE American control conference (ACC), pp 1610–1617
15. Alvarenga J, Vitzilaios NI, Valavanis KP, Rutherford MJ (2015) Survey of unmanned helicopter model-based navigation and control techniques. J Intell Robot Syst 80(1):87–138
16. Jetley P, Sujit PB, Saripalli S (2017) Safe landing of fixed wing UAVs. In: IEEE international conference on dependable systems and networks workshop (DSN-W), pp 2–9
17. Pandey AK, Chaudhary T, Mishra S, Verma S (2018) Longitudinal control of small unmanned aerial vehicle by PID controller. Intelligent communication, control and devices. Springer, Berlin, pp 923–931
18. Poksawat P, Wang L, Mohamed A (2017) Gain scheduled attitude control of fixed-wing UAV with automatic controller tuning. IEEE Trans Control Syst Technol
19. Espinoza-Fraire T, Dzul A, Cortes-Martinez F, Giernacki W (2018) Real-time implementation and flight tests using linear and nonlinear controllers for a fixed-wing miniature aerial vehicle (MAV). Int J Control, Autom Syst 16(1):392–396

20. Sarhan A, Qin S (2017) Robust adaptive flight controller for UAV systems. In: IEEE international conference on information science and control engineering (ICISCE), pp 1214–1219
21. Lyapunov AM (1992) The general problem of the stability of motion. Int J Control 55(3):531–534
22. Krstic M, Kanellakopoulos I, Kokotovic PV (1995) Nonlinear and adaptive control design. Wiley, Hoboken
23. Harkegard O (2003) Backstepping and control allocation with applications to flight control. PhD thesis, Linkopings universitet
24. Ren W, Atkins E (2005) Nonlinear trajectory tracking for fixed wing UAVs via backstepping and parameter adaptation. In: Proceedings of the AIAA guidance, navigation, and control conference and exhibit
25. Brezoescu A, Espinoza T, Castillo P, Lozano R (2013) Adaptive trajectory following for a fixed-wing UAV in presence of crosswind. J Intell Robot Syst
26. Yoon S, Kim Y, Park S (2012) Constrained adaptive backstepping controller design for aircraft landing in wind disturbance and actuator stuck. Int J Aeronaut Space Sci 13(1):74–89
27. Wang YC, Chen WS, Zhang SX, Zhu JW, Cao LJ (2018) Command-filtered incremental backstepping controller for small unmanned aerial vehicles. J Guid, Control, Dyn 41(4):954–967
28. Afandi MNRBM, Hassan MB, Suhardi G, Zhou Y, Danny L (2017) Comparison of backstepping, fuzzy-PID, and PID control techniques using X8 model in relation to A^* path planning. In: IEEE international conference on intelligent transportation engineering (ICITE), pp 340–345
29. Utkin V (2009) Sliding mode control. Nonlinear, distributed, and time delay systems, control, systems, robotics and automation
30. Vaidyanathan S, Lien CH (2017) Applications of sliding mode control in science and engineering. Springer, Berlin
31. Khalil HK (2014) Nonlinear control. Prentice Hall, Upper Saddle River
32. Castaneda H, Salas-Pena OS, de Leon-Morales J (2017) Extended observer based on adaptive second order sliding mode control for a fixed wing UAV. ISA Trans 66:226–232
33. Zheng Z, Jin Z, Sun L, Zhu M (2017) Adaptive sliding mode relative motion control for autonomous carrier landing of fixed-wing unmanned aerial vehicles. IEEE Access 5:5556–5565
34. Espinoza-Fraire AT, Chen Y, Dzul A, Lozano R, Juarez R (2018) Fixed-wing MAV adaptive PD control based on a modified MIT rule with sliding-mode control. J Intell Robot Syst
35. Lu B, Fang Y, Sun N (2018) Continuous sliding mode control strategy for a class of nonlinear underactuated systems. IEEE Trans Autom Control
36. Perozzi G, Efimov D, Biannic JM, Planckaert L, Coton P (2017) Wind rejection via quasi-continuous sliding mode technique to control safely a mini drone. In: European conference for aeronautics and space science
37. Rios H, Gonzalez-Sierra J, Dzul A (2017) Robust tracking output-control for a quad-rotor: a continuous sliding-mode approach. J Frankl Inst 354(15):6672–6691
38. Munoz-Vazquez AJ, Parra-Vega V, Sanchez-Orta A (2017) Continuous fractional-order sliding PI control for nonlinear systems subject to non-differentiable disturbances. Asian J Control 19(1):279–288
39. Espinoza T, Dzul AE, Lozano R, Parada P (2014) Backstepping-sliding mode controllers applied to a fixed-wing UAV. J Intell Robot Syst 73(1–4):67–79
40. Ru P, Subbarao K (2017) Nonlinear model predictive control for unmanned aerial vehicles. Aerospace 4(2)
41. Ulker H, Baykara C, Ozsoy C (2017) Design of MPCs for a fixed wing UAV. Aircr Eng Aerosp Technol 89(6):893–901
42. Eren U, Prach A, Kocer BB, Rakovic SV, Kayacan E, Acikmese B (2017) Model predictive control in aerospace systems: current state and opportunities. J Guid, Control, Dyn
43. Kang Y, Hedrick JK (2009) Linear tracking for a fixed-wing UAV using nonlinear model predictive control. IEEE Trans Control Syst Technol 17(5):1202–1210
44. Yang K, Kang Y, Sukkarieh S (2013) Adaptive nonlinear model predictive path-following control for a fixed-wing unmanned aerial vehicle. Int J Control, Autom Syst 11(1):65–74

45. Stastny TJ, Dash A, Siegwart R (2017) Nonlinear MPC for fixed-wing UAV trajectory tracking: implementation and flight experiments. In: AIAA guidance, navigation, and control conference
46. Gavilan F, Vazquez R, Lobato A, de la Rosa M, Gallego A, Camacho EF, Hardt MW, Navarro FA (2018) Increasing predictability and performance in UAS flight contingencies using AIDL and MPC. In: AIAA guidance, navigation, and control conference
47. Jain RPK, Aguiar AP, Alessandretti A, Borges de Sousa J (2018) Moving path following control of constrained underactuated vehicles: a nonlinear model predictive control approach. In: AIAA information systems - AIAA infotech aerospace
48. Astrom KJ, Wittenmark B (2013) Adaptive control. Courier Corporation
49. Bellman RE (2015) Adaptive control processes: a guided tour. Princeton University Press, Princeton
50. Gavilan F, Acosta JA, Vazquez R (2011) Control of the longitudinal flight dynamics of an UAV using adaptive backstepping. IFAC Proc Vol 44(1):1892–1897
51. Ambati PR, Padhi R (2017) Robust auto-landing of fixed-wing UAVs using neuro-adaptive design. IFAC Control Eng Pract 60:218–232
52. de Oliveira HA, Rosa PFF (2017) Adaptive genetic neuro-fuzzy attitude control for a fixed wing UAV. In: IEEE international conference on industrial technology (ICIT), pp 726–731
53. Noble D, Bhandari S (2017) Neural network based nonlinear model reference adaptive controller for an unmanned aerial vehicle. In: IEEE international conference on unmanned aircraft systems (ICUAS), pp 94–103
54. Ioannou PA, Sun J (2010) Robust adaptive control. Prentice-Hall, Upper Saddle River (Reprint)
55. Hespanha JP, Liberzon D, Morse AS (2003) Overcoming the limitations of adaptive control by means of logic-based switching. Syst Control Lett 49(1):49–65
56. Stefanovic M, Safonov MG (2011) Safe adaptive control: data-driven stability analysis and robust synthesis. Springer, Berlin
57. Long L, Zhao J (2017) Adaptive control for a class of high-order switched nonlinearly parameterized systems. Int J Robust Nonlinear Control 27(4):547–565
58. Wang H, Wang Z, Liu YJ, Tong S (2017) Fuzzy tracking adaptive control of discrete-time switched nonlinear systems. Fuzzy Sets Syst 316:35–48
59. Li Y, Sui S, Tong S (2017) Adaptive fuzzy control design for stochastic nonlinear switched systems with arbitrary switchings and unmodeled dynamics. IEEE Trans Cybern 47(2):403–414
60. Li Y, Tong S (2017) Adaptive fuzzy output-feedback stabilization control for a class of switched nonstrict-feedback nonlinear systems. IEEE Trans Cybern 47(4):1007–1016
61. Cho SJ, Lee JS, Kim J, Kuc TY, Chang PH, Jin M (2017) Adaptive time-delay control with a supervising switching technique for robot manipulators. Trans Inst Meas Control 39(9):1374–1382
62. Enns D, Bugajski D, Hendrick R, Stein G (1994) Dynamic inversion: an evolving methodology for flight control design. Int J Control 59(1):71–91
63. Kawakami Y, Uchiyama K (2017) Nonlinear controller design for transition flight of a fixed-wing UAV with input constraints. In: AIAA guidance, navigation, and control conference
64. Chang H, Liu Y, Wang Y, Zheng X (2017) A modified nonlinear dynamic inversion method for attitude control of UAVs under persistent disturbances. In: IEEE international conference on information and automation (ICIA), pp 715–721
65. Smeur EJ, de Croon GCHE, Chu Q (2018) Cascaded incremental nonlinear dynamic inversion for MAV disturbance rejection. Control Eng Pract 73:79–90
66. Amlashi AH, Gharamaleki RM, Nejad MHH, Mirzaei M (2018) Design of estimator-based nonlinear dynamic inversion controller and nonlinear regulator for robust trajectory tracking with aerial vehicles. Int J Dyn Control 6(2):707–725
67. Silva NB, Fontes JV, Inoue RS, Branco KR (2018) Dynamic inversion and gain-scheduling control for an autonomous aerial vehicle with multiple flight stages. J Control, Autom Electr Syst 29(3):328–339
68. Kurnaz S, Cetin O, Kaynak O (2009) Fuzzy logic based approach to design of flight control and navigation tasks for autonomous unmanned aerial vehicles. J Intell Robot Syst 54(1–3):229–244

69. Espinoza T, Dzul A, Llama M (2013) Linear and nonlinear controllers applied to fixed-wing UAV. Int J Adv Robot Syst 10(1)
70. Lee T, Kim Y (2001) Nonlinear adaptive flight control using backstepping and neural networks controller. J Guid, Control, Dyn 24(4):675–682
71. Kayacan E, Khanesar MA, Rubio-Hervas J, Reyhanoglu M (2017) Learning control of fixed-wing unmanned aerial vehicles using fuzzy neural networks. Int J Aerosp Eng
72. de Oliveira HA, Rosa PFF (2017) Genetic neuro-fuzzy approach for unmanned fixed wing attitude control. In: IEEE international conference on military technologies (ICMT), pp 485–492
73. Santoso F, Garratt MA, Anavatti SG (2018) State-of-the-art intelligent flight control systems in unmanned aerial vehicles. IEEE Trans Autom Sci Eng 15(2):613–627
74. Rugh WJ, Shamma JS (2000) Research on gain scheduling. Automatica 36(10):1401–1425
75. Marcos A, Balas GJ (2004) Development of linear-parameter-varying models for aircraft. J Guid Control Dyn 27(2):218–228
76. Girish CV, Emilio F, Jonathan HP, Hugh L (2015) Nonlinear flight control techniques for unmanned aerial vehicles. Handbook of unmanned aerial vehicles. Springer, Berlin, pp 577–612
77. Shao P, Zhou Z, Ma S, Bin L (2017) Structural robust gain-scheduled PID control and application on a morphing wing UAV. In: IEEE Chinese control conference (CCC), pp 3236–3241
78. Yue T, Wang L, Ai J (2017) Longitudinal integrated linear parameter varying control for morphing aircraft in large flight envelope. In: AIAA atmospheric flight mechanics conference
79. Stephan J, Fichter W (2018) Gain-scheduled multivariable flight control under uncertain trim conditions. In: AIAA guidance, navigation, and control conference
80. Petersen IR, Ugrinovskii VA, Savkin AV (2012) Robust control design using H_∞ methods. Springer Science & Business Media
81. Glover K (2015) H-infinity control. Encyclopedia of systems and control
82. Ferreira HC, Baptista RS, Ishihara JY, Borges GA (2011) Disturbance rejection in a fixed wing UAV using nonlinear H_∞ state feedback. In: IEEE international conference on control and automation (ICCA), pp 386–391
83. Lesprier J, Biannic JM, Roos C (2015) Modeling and robust nonlinear control of a fixed-wing UAV. In: IEEE conference on control applications (CCA), pp 1334–1339
84. Biannic JM, Roos C, Lesprier J (2017) Nonlinear structured H_∞ controllers for parameter-dependent uncertain systems with application to aircraft landing. AerospaceLab J
85. Kung CC (2008) Nonlinear H_∞ robust control applied to F-16 aircraft with mass uncertainty using control surface inverse algorithm. J Frankl Inst 345(8):851–876
86. Garcia GA, Kashmiri S, Shukla D (2017) Nonlinear control based on H-infinity theory for autonomous aerial vehicle. In: IEEE international conference on unmanned aircraft systems (ICUAS), pp 336–345
87. Balas GJ, Doyle JC, Glover K, Packard A, Smith R (1993) μ-analysis and synthesis toolbox. MUSYN Inc. and The MathWorks
88. Balas G, Chiang R, Packard A, Safonov M (2005) Robust control toolbox, for use with matlab. User's Guide
89. Yin G, Chen N, Li P (2007) Improving handling stability performance of four-wheel steering vehicle via μ-synthesis robust control. IEEE Trans Veh Technol 56(5):2432–2439

90. Qiu L, Fan G, Yi J, Yu W (2009) Robust hybrid Controller design based on feedback lineariza-
    tion and μ-synthesis for UAV. In: IEEE international conference on intelligent computation
    technology and automation (ICICTA), pp 858–861
91. Michailidis MG, Kanistras K, Agha M, Rutherford MJ, Valavanis KP (2018) Nonlinear control
    of fixed-wing UAVs with time-varying aerodynamic uncertainties via μ-synthesis. In: IEEE
    conference on decision and control (CDC), pp 6314–6321

Chapter 3
Technical Problem Statement

Abstract This chapter presents necessary background related to UAV motion during flight, answering the fundamental question of what is the source of aerodynamic uncertainty through the aircraft equations of motion and the detailed breakdown of the aerodynamic coefficients. The basic UAV reference frames are presented, the control surfaces are introduced and the direct relation between the equations of motion and the aircraft aerodynamic coefficients is established. Finally, potential and existing real-life applications of UAVs with inherent uncertain aerodynamic parameters are demonstrated.

3.1 Basic Reference Frames

Inertial frame (Fig. 3.1) and vehicle frame (Fig. 3.2) are usually referred to as NED frames (north, east and down) and the difference between these two is that the origin of the vehicle frame is located to the center of gravity (CoG) of the aircraft.

Vehicle-1 (Fig. 3.3), vehicle-2 (Fig. 3.4) and body (Fig. 3.5) frames are used to define the basic attitude angles of the UAV (also known as Euler angles), pitch θ, roll ϕ and heading (yaw) ψ. Angle of attack α, sideslip β, flight path γ and course angle χ are defined respectively by the frames of Figs. 3.6, 3.7 and 3.8.

For the stability frame, i^s axis is pointing along the projection of the airspeed vector onto the $i^b - k^b$ plane of the body frame, j^s axis is the same as j^b axis of the body frame and k^s is formed so that a right-handed coordinate system is designed. For the wind frame, i^w axis points along the direction of the airspeed vector.

Flight path angle γ is the angle between the horizontal plane and the ground velocity vector V_g, while course angle χ is the angle between the projection of the ground velocity onto the horizontal plane and true north.

© Springer Nature Switzerland AG 2020   29
M. G. Michailidis et al., *Nonlinear Control of Fixed-Wing UAVs*
with Time-Varying and Unstructured Uncertainties, Springer Tracts
in Autonomous Systems 1, https://doi.org/10.1007/978-3-030-40716-2_3

Fig. 3.1 The inertial frame
[1]

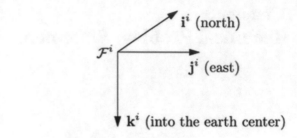

Fig. 3.2 The vehicle frame
[1]

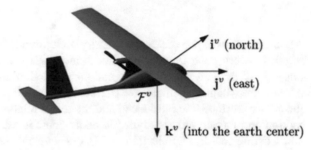

Fig. 3.3 The vehicle-1
frame [1]

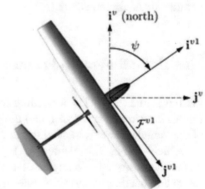

3.2 Equations of Motion

The UAV main axes of motion (pitch, roll and yaw) are shown in Fig. 3.9, which also defines the UAV body frame $F^b = (i^b, j^b, k^b)$.

Table 3.1 gives the state variables for the UAV equations of motion. The NED positions of the UAV (p_n, p_e, p_d) are defined relative to the earth reference frame while the linear and angular velocities are defined with respect to the body frame F^b.

Fig. 3.4 The vehicle-2
frame [1]

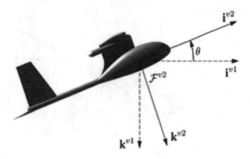

Fig. 3.5 The body frame [1]

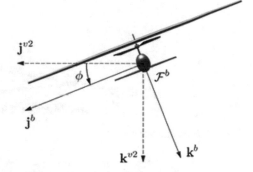

Fig. 3.6 The stability frame
[1]

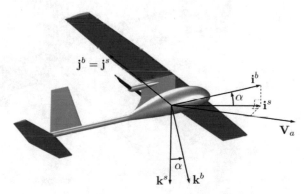

Fig. 3.7 The wind frame [1]

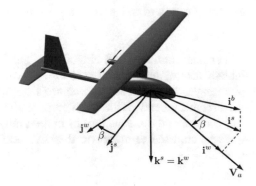

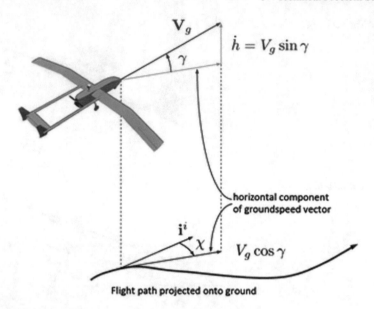

Fig. 3.8 Flight path angle [1]

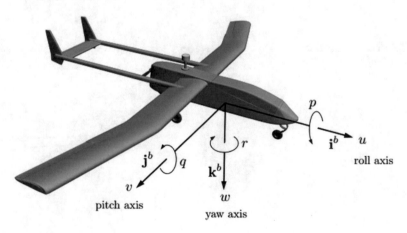

Fig. 3.9 UAV axes of motion [1]

For conventional T-tail aircraft configurations (Fig. 3.10), the control surfaces include the aileron δ_a, used to control the roll angle, the elevator δ_e that affects the pitch angle, the rudder δ_r which controls the yaw angle and the throttle δ_t, controlling the aircraft's speed V_a. Other configurations include V-tail, flying wing, spoilers, flaps and canards and the equations of motion may be defined accordingly. The control surfaces are used to maneuver the UAV and modify the aerodynamic forces and moments.

Table 3.1 State variables for UAV equations of motion [1]

Variable	Description
p_n	Inertial north position of the UAV
p_e	Inertial east position of the UAV
p_d	Inertial down position (negative of altitude)
u	Velocity along i^b
v	Velocity along j^b
w	Velocity along k^b
ϕ	Roll angle of the UAV
θ	Pitch angle of the UAV
ψ	Heading (yaw) angle of the UAV
p	Roll rate
q	Pitch rate
r	Yaw rate

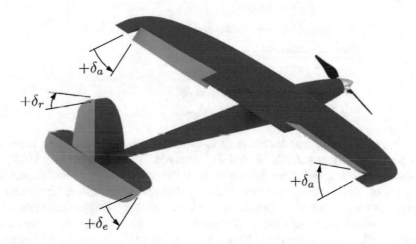

Fig. 3.10 T-tail UAV configuration [1]

The complete set of the navigation, force, kinematic and moment equations that govern the dynamic behavior of the UAV during flight, as found in [1, 2], is given in (3.1).

Navigation equations

$$\dot{p}_n = (\cos\theta\cos\psi)u + (\sin\phi\sin\theta\cos\psi - \cos\phi\sin\psi)v$$
$$+ (\cos\phi\sin\theta\cos\psi + \sin\phi\sin\psi)w$$
$$\dot{p}_e = (\cos\theta\sin\psi)u + (\sin\phi\sin\theta\sin\psi + \cos\phi\cos\psi)v$$
$$+ (\cos\phi\sin\theta\sin\psi - \sin\phi\cos\psi)w$$
$$\dot{p}_d = u\sin\theta - v\sin\phi\cos\theta - w\cos\phi\cos\theta$$

Force equations

$$\dot{u} = rv - qw - g\sin\theta + F_{i^b}/m$$
$$\dot{v} = pw - ru + g\cos\theta\sin\phi + F_{j^b}/m$$
$$\dot{w} = qu - pv + g\cos\theta\cos\phi + F_{k^b}/m \qquad\qquad (3.1)$$

Kinematic equations

$$\dot{\phi} = p + q\sin\phi\tan\theta + r\cos\phi\tan\theta$$
$$\dot{\theta} = q\cos\phi - r\sin\phi$$
$$\dot{\psi} = q\sin\phi\sec\theta + r\cos\phi\sec\theta$$

Moment equations

$$\dot{p} = \Gamma_1 pq - \Gamma_2 qr + \Gamma_3\ell + \Gamma_4 n$$
$$\dot{q} = \Gamma_5 pr - \Gamma_6(p^2 - r^2) + m/J_y$$
$$\dot{r} = \Gamma_7 pq - \Gamma_1 qr + \Gamma_4\ell + \Gamma_8 n$$

The parameters Γ_i that appear in (3.1) are functions of the aircraft's moments and products of inertia J_x, J_y, J_z and J_{xz}, analytically defined in (3.2). UAV is a nonlinear underactuated dynamic system, with its motion mathematically described by a set of 12 coupled, first-order, ordinary differential equations. The standard convention for the dynamics of a fixed-wing aircraft is an approximate decomposition into longitudinal and lateral motion. While there is coupling between the two motions, for most airframes this dynamic effect is sufficiently small and can be mitigated by control algorithms for disturbance rejection.

$$\Gamma = J_x J_z - J_{xz}^2$$

$$\Gamma_1 = \frac{J_{xz}(J_x - J_y + J_z)}{\Gamma}$$

$$\Gamma_2 = \frac{J_z(J_z - J_y) + J_{xz}^2}{\Gamma}$$

$$\Gamma_3 = \frac{J_z}{\Gamma}$$

$$\Gamma_4 = \frac{J_{xz}}{\Gamma} \qquad\qquad (3.2)$$

Table 3.2 UAV aerodynamic forces and moments [1]

Parameter	Description
F_{i^b}	Total force along i^b axis
F_{j^b}	Total force along j^b axis
F_{k^b}	Total force along k^b axis
ℓ	Rolling moment
m	Pitching moment
n	Yawing moment
C_L	Lift coefficient
C_D	Drag coefficient
C_Y	Sideforce coefficient
C_l	Rolling moment coefficient
C_m	Pitching moment coefficient
C_n	Yawing moment coefficient

$$\Gamma_5 = \frac{J_z - J_x}{J_y}$$

$$\Gamma_6 = \frac{J_{xz}}{J_y}$$

$$\Gamma_7 = \frac{(J_x - J_y)J_x + J_{xz}^2}{\Gamma}$$

$$\Gamma_8 = \frac{J_x}{\Gamma}$$

3.3 Aerodynamic Uncertainty in the UAV Model

The aerodynamic forces and moments and their respective coefficients (Table 3.2) have a complex dependence on a large number of variables and this creates both modeling and measurement problems. Therefore, it is advantageous to build up an aerodynamic coefficient from a sum of components that provide physical insight and are convenient to handle mathematically.

The forces and moments acting on the airframe have a nonlinear dependence on the aerodynamic coefficients and the control surfaces that is established in (3.3).

$$F_{i^b} = F_{i^b}(C_L, C_D, \delta_e)$$
$$F_{j^b} = F_{j^b}(C_Y, \delta_\alpha, \delta_r)$$
$$F_{k^b} = F_{k^b}(C_L, C_D, \delta_e)$$
$$\ell = \ell(C_l, \delta_\alpha, \delta_r) \tag{3.3}$$
$$m = m(C_m, \delta_e)$$
$$n = n(C_n, \delta_\alpha, \delta_r)$$

In general, the aerodynamic coefficients appearing in (3.3) are nonlinear. Nevertheless, they can be modeled with acceptable accuracy using linear approximations like Taylor series expansion. Making use of this approach, the coefficients can be simplified for better handling as in (3.4).

$$C_L = C_L(C_{L_0}, C_{L_\alpha}, C_{L_q}, C_{L_{\delta_e}})$$
$$C_D = C_D(C_{D_0}, C_{D_\alpha}, C_{D_q}, C_{D_{\delta_e}})$$
$$C_Y = C_Y(C_{Y_0}, C_{Y_\beta}, C_{Y_p}, C_{Y_r}, C_{Y_{\delta_\alpha}}, C_{Y_{\delta_r}})$$
$$C_l = C_l(C_{l_0}, C_{l_\beta}, C_{l_p}, C_{l_r}, C_{l_{\delta_\alpha}}, C_{l_{\delta_r}}) \tag{3.4}$$
$$C_m = C_m(C_{m_0}, C_{m_\alpha}, C_{m_q}, C_{m_{\delta_e}})$$
$$C_n = C_n(C_{n_0}, C_{n_\beta}, C_{n_p}, C_{n_r}, C_{n_{\delta_\alpha}}, C_{n_{\delta_r}})$$

where α is the angle of attack, β is the sideslip angle and the subscript 0 is the value of the respective coefficient when the linearizing variables are set to 0. The terms inside the parentheses on the right hand side of (3.4) are dimensionless quantities called control and stability derivatives.

The label derivative comes from the fact that the coefficients originated as partial derivatives in the Taylor series approximation. Having established (3.3) and (3.4), the connection between aerodynamic changes/uncertainty and the impact on the UAV model is clear. Aerodynamic uncertainty is by default present in the UAV model for controller design purposes due to the challenging task of accurate estimation of the control and stability derivatives. In addition to that, any aerodynamic changes on the UAV can be reflected on the aircraft control and stability derivatives.

3.4   UAVs with Uncertain Aerodynamic Parameters

Focusing on the UAV control and stability derivatives, this section presents potential or existing real-life applications of new generation UAVs where this study may be immediately helpful for researchers.

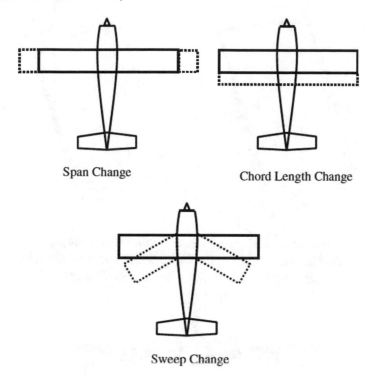

Span Change   Chord Length Change

Sweep Change

Fig. 3.11   In-plane shape morphing [7]

3.4.1   Morphing Aircraft

It is always fascinating when technology mimics nature. Motivated by the Greek word *"morpho"*, which means to transform, morphing technology seeks to emulate the biological structure of a bird [3]. This new class of UAVs will be able to control itself like a bird, with wings that twist, fold and transform [4, 5]. Morphing research projects such as the MFX-1 developed by NextGen Aeronautics [6] will revolutionize the costs of building and operating aircraft.

The objective of morphing technologies is to develop high performance aircraft with wings designed to change shape and performance substantially during flight to create multiple-regime, aerodynamically-efficient, shape-changing aircraft. The morphing wing change of shape can occur either in-plane (Fig. 3.11), or out-of-plane as shown in Fig. 3.12.

For a detailed definition of the airfoil characteristics the reader is referred to [8]. Morphing technologies will be used to improve aircraft performance, make them more efficient and enable the vehicles to operate under a wide range of flight conditions.

Morphing UAVs belong to the general class of active wing shaping UAVs which enable complex trailing-edge shapes that could contribute to aerodynamic, structural,

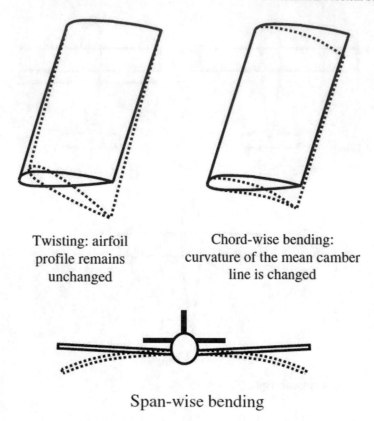

Twisting: airfoil   Chord-wise bending:
profile remains   curvature of the mean camber
unchanged   line is changed

Span-wise bending

Fig. 3.12 Out-of-plane shape morphing [7]

and control advantages. An example can be seen in Fig. 3.13, which shows a UAV with
segmented control surfaces for improved aerodynamic efficiency. Changing shape
during flight implies an on-demand alteration of all the aircraft aerodynamic characteristics. Therefore, this study will be a useful tool for those conducting research
in the field of navigation and control of morphing and active wing shaping UAVs.

3.4.2 Delivery UAVs

Interest in small flying machines as means of delivering payloads has been continuously increasing and the idea of turning UAVs into a commercialized delivery
mechanism has sparked a lot of debate. Some of the numerous applications include
delivery of food products, providing assistance in the agricultural and farming industry, supply chain applications, package delivering and last but not least, the use of
UAVs for medical purposes [10, 11]. Amazon, Google and UPS are some of the

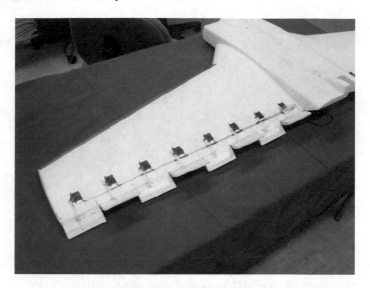

Fig. 3.13 Segmented left wing deflected to induce heading moment [9]

industry leaders that have initiated research on new, UAV-assisted product-delivery methods.

For a technical justification of the importance of this study when it comes to UAVs as delivery mechanisms, the relation between the control and stability derivatives and the vehicle's mass must be identified. The traditional approach for aircraft system identification is the derivation of a linear model based on motion decoupling [12]. For the UAV lateral and longitudinal motion, a state-space model $\dot{x} = Ax + Bu$ is derived through flight testing. The elements of matrix A are functions of the aircraft's control and stability derivatives and trim flight conditions. An example can be seen in (3.5), which gives the actual mathematical expression for the lateral state-space model coefficient Y_r as a function of the aircraft's mass m.

$$Y_r = -u^* + \frac{\rho\, V_\alpha^*\, S\, b}{4m} C_{Y_r} \tag{3.5}$$

Nomenclature and the complete tables of the lateral and longitudinal state-space model coefficients can be found in [1]. As an example, the relation between lateral state-space coefficients and the lateral control and stability derivatives is illustrated in Fig. 3.14.

Assuming changing mass due to a delivery scenario means that the UAV control and stability derivatives (parameter on the left of Fig. 3.14) will have time-varying values during flight as (3.5) dictates. Hence, the controller design methodology that will be applied on a delivery UAV, will inevitably be one of the techniques investigated in this research monograph.

Lateral	Formula
Y_v	$\frac{\rho S b v^*}{4 m V_a^*}\left[C_{Y_p} p^* + C_{Y_r} r^*\right] + \frac{\rho S v^*}{m}\left[C_{Y_0} + C_{Y_\beta}\beta^* + C_{Y_{\delta_a}}\delta_a^* + C_{Y_{\delta_r}}\delta_r^*\right] + \frac{\rho S C_{Y_\beta}}{2m}\sqrt{u^{*2}+w^{*2}}$
Y_p	$w^* + \frac{\rho V_a^* S b}{4m} C_{Y_p}$
Y_r	$-u^* + \frac{\rho V_a^* S b}{4m} C_{Y_r}$
Y_{δ_a}	$\frac{\rho V_a^{*2} S}{2m} C_{Y_{\delta_a}}$
Y_{δ_r}	$\frac{\rho V_a^{*2} S}{2m} C_{Y_{\delta_r}}$
L_v	$\frac{\rho S b^2 v^*}{4 V_a^*}\left[C_{p_p} p^* + C_{p_r} r^*\right] + \rho S b v^*\left[C_{p_0} + C_{p_\beta}\beta^* + C_{p_{\delta_a}}\delta_a^* + C_{p_{\delta_r}}\delta_r^*\right] + \frac{\rho S b C_{p_\beta}}{2}\sqrt{u^{*2}+w^{*2}}$
L_p	$\Gamma_1 q^* + \frac{\rho V_a^* S b^2}{4} C_{p_p}$
L_r	$-\Gamma_2 q^* + \frac{\rho V_a^* S b^2}{4} C_{p_r}$
L_{δ_a}	$\frac{\rho V_a^{*2} S b}{2} C_{p_{\delta_a}}$
L_{δ_r}	$\frac{\rho V_a^{*2} S b}{2} C_{p_{\delta_r}}$
N_v	$\frac{\rho S b^2 v^*}{4 V_a^*}\left[C_{r_p} p^* + C_{r_r} r^*\right] + \rho S b v^*\left[C_{r_0} + C_{r_\beta}\beta^* + C_{r_{\delta_a}}\delta_a^* + C_{r_{\delta_r}}\delta_r^*\right] + \frac{\rho S b C_{r_\beta}}{2}\sqrt{u^{*2}+w^{*2}}$
N_p	$\Gamma_7 q^* + \frac{\rho V_a^* S b^2}{4} C_{r_p}$
N_r	$-\Gamma_1 q^* + \frac{\rho V_a^* S b^2}{4} C_{r_r}$
N_{δ_a}	$\frac{\rho V_a^{*2} S b}{2} C_{r_{\delta_a}}$
N_{δ_r}	$\frac{\rho V_a^{*2} S b}{2} C_{r_{\delta_r}}$

Fig. 3.14 Mathematical definition of lateral state-space coefficients [1]

3.5　Remarks

Important background information and a technical problem statement of the research problem under investigation have been covered in this chapter. The goal is to highlight the dependence of changing, time-varying aerodynamic parameters on the mathematical model of the UAV during flight. Up next comes a complete and detailed analysis of the controller design algorithm based on the aerodynamic coefficients and the respective mathematical expressions that were given in this chapter.

References

1. Beard RW, McLain TW (2012) Small unmanned aircraft: theory and practice. Princeton University Press, Princeton
2. Stevens BL, Lewis FL, Johnson EN (2015) Aircraft control and simulation: dynamics, controls design, and autonomous systems. Wiley, New York
3. Rodriguez AR (2007) Morphing aircraft technology survey. In: 45th AIAA aerospace sciences meeting and exhibit
4. Barbarino S, Bilgen O, Ajaj RM, Friswell MI, Inman DJ (2011) A review of morphing aircraft. J Intell Mater Syst Struct 22(9):823–877
5. Tavares SMO, Moreira SJ, de Castro PMST, Gamboa PV (2017) Morphing aeronautical structures: a review focused on UAVs and durability assessment. In: IEEE international conference on actual problems of unmanned aerial vehicles developments (APUAVD), pp 49–52
6. Flanagan JS, Strutzenberg RC, Myers RB, Rodrian JE (2007) Development and flight testing of a morphing aircraft, the NextGen MFX-1
7. Sofla AYN, Meguid SA, Tan KT, Yeo WK (2010) Shape morphing of aircraft wing: status and challenges. Mater Des 31(3):1284–1292

8. Anderson JD (2010) Fundamentals of aerodynamics. McGraw-Hill Education, New York
9. Abdulrahim M (2003) Flight dynamics and control of an aircraft with segmented control surfaces. In: 42nd AIAA aerospace sciences meeting and exhibit
10. Mo J, Chen AZ (2017) UAV delivery system design and analysis. In: 17th Australian international aerospace congress (AIAC)
11. Erdelj M, Natalizio E, Chowdhury KR, Akyildiz IF (2017) Help from the sky: leveraging UAVs for disaster management. IEEE Pervasive Comput 16(1):24–32
12. Klein V, Morelli EA (2006) Aircraft system identification: theory and practice. American Institute of Aeronautics and Astronautics, Reston

Chapter 4
Controller Design

Abstract Flight control design for unconventional UAVs with time-varying aerodynamic uncertainties focuses particularly on how the uncertain parameters can be incorporated in the aircraft mathematical model. Lift, drag, sideforce, rolling moment, pitching moment and heading moment coefficients must be appropriately modeled so that uncertainty can be tackled. Literature has already shown that the exact calculation of the aerodynamic characteristics of a specific aircraft is a challenging task [1, 2]. This is why estimations and computer software such as XFLR5, found in [3], are employed to derive approximations. The solution considered to overcome this obstacle is μ-analysis of uncertain systems and additive uncertainty weighting functions. Before presenting the complete process for the proposed UAV controller framework, there is the need to demonstrate how uncertainty can be represented on a given system for μ-synthesis to be applied.

4.1 μ-Synthesis Preliminaries

The basic idea in modeling an uncertain system is to separate what is known from what is unknown in a feedback-like connection, and bound the possible values of the unknown elements. As explained in [4], the theory for representing uncertainty in matrices and systems is Linear Fractional Transformations (LFTs). Suppose a transfer matrix M is given, relating input r and output v as in Fig. 4.1.

If r and v are partitioned into a top and bottom part, that is if M is a 2×2 transfer matrix, then the relationship can be drawn in more detail, explicitly showing the partitioned matrix M as shown in Fig. 4.2.

Suppose now a matrix Δ relates r_2 to v_2 as in Fig. 4.3.

The LFT of M by Δ interconnects these two elements as shown in Fig. 4.4.

Eliminating r_2 and v_2 yields:

$$v_1 = F_L(M, \Delta)r_1 = [M_{11} + M_{12}\Delta(I - M_{22}\Delta)^{-1}M_{21}]r_1 \qquad (4.1)$$

The lower loop of Fig. 4.4 can be considered as the uncertainty block of the initial system. The notation $F_L(M, \Delta)$ is used because the Δ block is on the lower loop of

© Springer Nature Switzerland AG 2020   43
M. G. Michailidis et al., *Nonlinear Control of Fixed-Wing UAVs*
with Time-Varying and Unstructured Uncertainties, Springer Tracts
in Autonomous Systems 1, https://doi.org/10.1007/978-3-030-40716-2_4

$$v = Mr$$

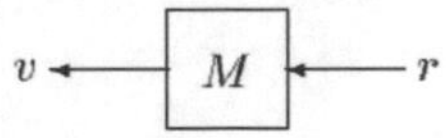

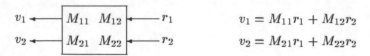

Fig. 4.1 Transfer matrix M relating r and v [4]

$$v_1 = M_{11}r_1 + M_{12}r_2$$
$$v_2 = M_{21}r_1 + M_{22}r_2$$

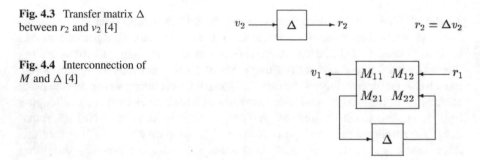

Fig. 4.2 Partitioned transfer matrix M [4]

Fig. 4.3 Transfer matrix Δ between r_2 and v_2 [4]

$$r_2 = \Delta v_2$$

Fig. 4.4 Interconnection of M and Δ [4]

Fig. 4.4. Inserting the Δ block in the upper loop of the interconnection and using the upper loop notation yields:

$$v_2 = F_U(M, \Delta)r_2 = [M_{22} + M_{21}\Delta(I - M_{11}\Delta)^{-1}M_{12}]r_2 \tag{4.2}$$

This is the theory needed to represent and analyze model uncertainty. Model uncertainty is divided into two categories: *Parametric* and *Multiplicative*. Parametric uncertainty refers to systems with uncertain parameters that have a nominal value and a range of possible variation. For instance, consider the second order spring system:

$$m\ddot{x} + c\dot{x} + kx = u \tag{4.3}$$

with uncertain parameters m, c, k. 50% uncertainty in m, 30% in c and 40% in k is considered. Let's also denote the nominal values for m, c, k as $\bar{m}, \bar{c}, \bar{k}$ respectively. This is translated into:

$$m = \bar{m}(1 + 0.5\delta_m), \quad c = \bar{c}(1 + 0.3\delta_c), \quad k = \bar{k}(1 + 0.4\delta_k) \tag{4.4}$$

The usual block diagram that would represent a known system as described in (4.3) is depicted in Fig. 4.5.

If uncertainties for m, c, k exist in the model, then by using the block diagrams seen in Fig. 4.6, the block diagram of Fig. 4.5 is converted into a new one including the uncertainties as shown in Fig. 4.7.

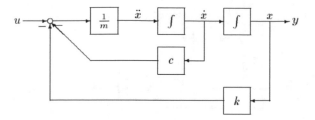

Fig. 4.5 Block diagram of a known 2nd order spring system [4]

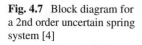

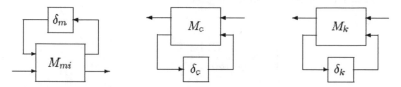

Fig. 4.6 Block diagrams representing uncertainties for m, c, k [4]

Fig. 4.7 Block diagram for a 2nd order uncertain spring system [4]

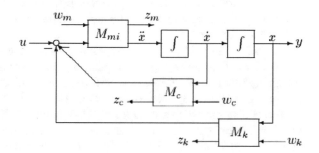

Although this type of uncertainty looks like it applies for a non-conventional UAV, it will not be effective. Parametric uncertainty is mainly aiming to deal with uncertainties stemming from a specific parameter itself. In other words, parametric uncertainty is usually applicable when the uncertainty exists in the model in a linear like fashion and can be distinguished from the known part or the measured variables. The UAV mathematical model is too complicated and highly nonlinear to be handled this way. And for the design of a robust nonlinear controller, we need something richer and more powerful to analyze it i.e., multiplicative uncertainty. Multiplicative uncertainty roughly allows one to specify a frequency-dependent percentage uncertainty in the actual plant behavior. For this two components are needed:

- A nominal plant model $G(s)$
- A multiplicative uncertainty weighting function $W_u(s)$

The precise definition of the multiplicative uncertainty weighting function is given by inequality (4.5).

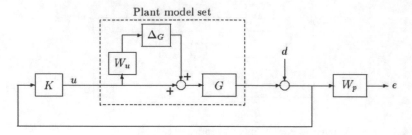

Fig. 4.8 Block diagram for a SISO uncertain plant [4]

$$\left| \frac{\tilde{G}(j\omega) - G(j\omega)}{G(j\omega)} \right| \leq |W_u(s)| \tag{4.5}$$

where $\tilde{G}(j\omega)$ denotes the overall uncertain transfer matrix. For a SISO uncertain plant, where the aircraft uncertainty is modeled as a complex full-block multiplicative uncertainty at the input of the rigid body aircraft nominal model, a block diagram as in Fig. 4.8 is sought.

The perturbation block Δ_G is supposed to satisfy the norm condition $\|\Delta_G\| \leq 1$. In the block diagram of Fig. 4.8, a plant output disturbance d and a performance weighting function W_p for certain performance tracking adjustments are also considered. The open loop interconnection of Fig. 4.8 is the one is sought regarding control of a single state variable for the UAV's motion. The design can be extended to a 2×2 transfer matrix (plant) as seen in Fig. 4.9, where G represents the uncertain plant model set as in Fig. 4.8. W_{del} functions denote actuator models and W_n functions are used to incorporate sensor noise. The structure of a generalized block diagram for a $n \times n$ MIMO uncertain plant is the same as Fig. 4.8 with appropriately adjusted dimensions for controller K, nominal model G, uncertainty weighting function W_u, perturbation block Δ_G, output disturbance d, performance weighting function W_p and control input u. Actuator and sensor models can also be introduced as in Fig. 4.9.

4.2   UAV Modeling and Uncertainty Range

Using Taylor series approximation as explained in [1, 2, 5], Eq. (4.6) presents the proposed aerodynamic coefficient modeling to realize the flight dynamics of a UAV with time-varying aerodynamic uncertainties.

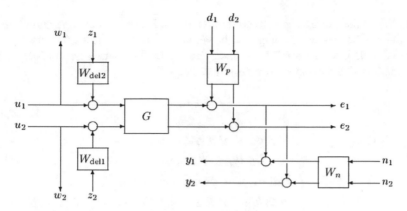

Fig. 4.9 Block diagram for a 2×2 MIMO uncertain plant [4]

$$C_L = C_{L_0}(\chi) + C_{L_\alpha}(\chi)\alpha + C_{L_q}(\chi)q + C_{L_{\delta_e}}(\chi)\delta_e$$
$$C_D = C_{D_0}(\chi) + \frac{(C_{L_0}(\chi) + C_{L_\alpha}(\chi)\alpha)^2}{\pi e AR}$$
$$C_Y = C_{Y_0}(\chi) + C_{Y_\beta}(\chi)\beta + C_{Y_p}(\chi)p + C_{Y_r}(\chi)r \qquad (4.6)$$
$$+ C_{Y_{\delta_\alpha}}(\chi)\delta_\alpha + C_{Y_{\delta_r}}(\chi)\delta_r$$
$$C_p = C_{p_0}(\chi) + C_{p_\beta}(\chi)\beta + C_{p_p}(\chi)p + C_{p_r}(\chi)r$$
$$+ C_{p_{\delta_\alpha}}(\chi)\delta_\alpha + C_{p_{\delta_r}}(\chi)\delta_r$$

For a detailed definition of the parameters that appear in Eq. (4.6) the reader is referred to Chap. 3 and [1]. Coefficients C_L, C_D, C_Y correspond to lift, drag and sideforce respectively while coefficient C_p can be broken down to two components, one for the rolling and one for the heading moment of the UAV. For instance:

$$C_{p_0} = \Gamma_3 C_{l_0} + \Gamma_4 C_{n_0} \qquad (4.7)$$

The first observation upon examining Eq. (4.6) is the dependence of the control and stability derivatives on an uncertain parameter χ, representing the respective uncertain aerodynamic configuration occurring at a given time instant. Thus, the control and stability derivatives can be represented as time-varying functions of χ. This slight modification provides an accurate theoretical modeling of the aerodynamic coefficients of a non-conventional UAV. For the time-varying control and stability derivatives, the assumption shown in Eq. (4.8) is used.

$$C_{L_{0_1}} \le C_{L_0}(\chi) \le C_{L_{0_2}} \qquad (4.8)$$

This inequality is applied for every time-varying derivative in Eq. (4.6). The quantity on the left, in this case $C_{L_{0_1}}$, is the value of $C_{L_0}(\chi)$ corresponding to the nominal model of the system, while the quantity on the right is the value of $C_{L_0}(\chi)$ corresponding to the extreme case, maximum value. This provides the operating interval and defines the uncertainty range for the vehicle's aerodynamic coefficients that will be used to define the overall UAV uncertain model for control system design. The assumptions listed in Eq. (4.9) are also considered.

$$V_a \leq V_{a_{max}}$$
$$\rho \leq \rho_{max}$$
$$\alpha \leq \alpha_{max}, \; \gamma \leq \gamma_{max}, \; \beta \leq \beta_{max}$$
$$\theta \leq \theta_{max}, \; \phi \leq \phi_{max}$$
$$\delta_e \leq \delta_{e_{max}}, \; \delta_\alpha \leq \delta_{\alpha_{max}}, \; \delta_r \leq \delta_{r_{max}} \qquad (4.9)$$
$$q \leq q_{max}, \; p \leq p_{max}, \; r \leq r_{max}$$
$$F_T \leq F_{T_{max}}$$

Maximum values regarding physical and control input constraints are considered for airspeed V_a, air density ρ, angle of attack α, flight path angle γ, sideslip angle β, pitch angle θ, roll angle ϕ, elevator deflection δ_e, aileron deflection δ_α, rudder δ_r pitch rate q, roll rate p, heading rate r and aircraft engine thrust F_T.

The inequality in Eq. (4.8) is the core of this monograph. The lower and upper bounds of Eq. (4.8) can be derived through aircraft system identification and flight testing for the respective lower and upper bound configurations. In the case of the UC^2AV, this implies aircraft system identification for the CC-off and CC-on cases. The state-space matrices A, B of $\dot{x} = Ax + Bu$ are identified for both the longitudinal and lateral motions. The dimensional control and stability derivatives can then be converted into the non-dimensional ones (as Eq. (4.8) dictates) through the longitudinal/lateral state-space coefficient tables that can be found in [1].

Another alternative that the authors implement and suggest is based on the foundation of adaptive control [6] for estimating the lower and upper bounds. A conceptual example of this method is given in the following equations, applied for the UC^2AV. The Dutch-Roll reduced order mode of the UAV is given by Eq. (4.10).

$$\begin{pmatrix} \dot{\bar{\beta}} \\ \dot{\bar{r}} \end{pmatrix} = \begin{pmatrix} Y_u & \frac{Y_r}{V_\alpha^* \cos \beta^*} \\ N_u V_\alpha^* \cos \beta^* & N_r \end{pmatrix} \begin{pmatrix} \bar{\beta} \\ \bar{r} \end{pmatrix} + \begin{pmatrix} \frac{Y_{\delta_r}}{V_\alpha^* \cos \beta^*} \\ N_{\delta_r} \end{pmatrix} \bar{\delta}_r \qquad (4.10)$$

where $Y_u, Y_r, Y_{\delta_r}, N_u, N_r, N_{\delta_r}$ are dimensional control and stability derivatives to be estimated, star notation denotes aircraft trim conditions and bar notation represents deviation from trim conditions. The UC^2AV has an approximate airspeed trim value of 21 m/s and a relatively small ($\approx$0) sideslip angle is considered. An accurate model of the UC^2AV has been designed using the XPlane flight simulator plane maker and a UDP communication between XPlane and Simulink has been established as described in [7]. The goal is to isolate the first differential equation, put it into linear parametric form (LPM), and employ an adaptive law that will estimate the unknown parameters through flight simulation. The LPM in this case can be written as in Eq. (4.11).

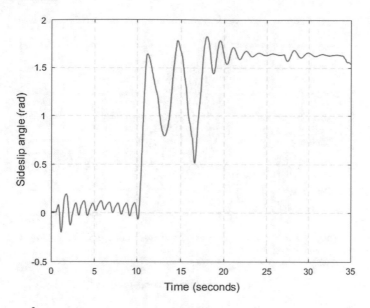

Fig. 4.10 UC^2AV sideslip response in XPlane

$$z = \theta\phi \;\Leftrightarrow\; \dot{\bar{\beta}} = \left(Y_u \; Y_r \; Y_{\delta_r} \right) \begin{pmatrix} \bar{\beta} \\ \bar{r}/21 \\ \bar{\delta}_r/21 \end{pmatrix} \qquad (4.11)$$

The rudder control input signal needs to be sufficiently rich, i.e., to have at least two fundamental frequencies, for persistence of excitation of the regressor signal ϕ to hold [8]. Therefore, the UC^2AV is excited by an automated rudder command generated by Simulink that is given by Eq. (4.12) in radians.

$$\delta_r(t) = 0.3(\sin 2\pi t + \sin 4\pi t) \qquad (4.12)$$

The sideslip and heading rate responses of the UC^2AV can be seen in Figs. 4.10 and 4.11 respectively. The control input is an automated signal and there is no human interference during the sweeps.

As a result, the aircraft loses stability after 10 s. The response signals are isolated from 0 to 10 s and converted into an approximate linear chirp waveform, with a particular emphasis on the amplitude and the frequency. For example, the sideslip response in radians is approximated by the waveform $\beta(t) = 0.17e^{-0.08t}\sin(-1.7\pi t)$ and the heading rate by the waveform $r(t) = 0.015e^{-0.08t}\sin(-1.7\pi t)$.

The unknown parameters of Eq. (4.11) are then estimated by the adaptive law $\dot{\hat{\theta}} = \Gamma\epsilon\phi$, with the tracking error ϵ given by $\epsilon = z - \hat{\theta}^T\phi$ and the diagonal positive

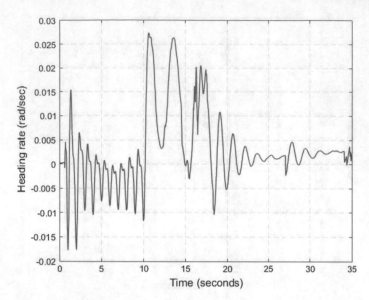

Fig. 4.11 UC^2AV heading rate response in XPlane

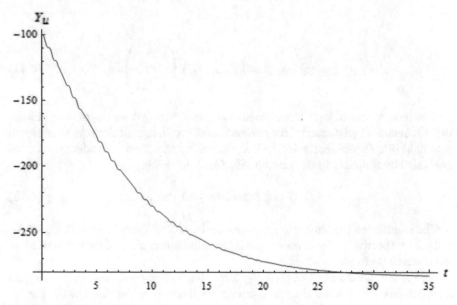

Fig. 4.12 Evolution of Y_u over time

definite matrix Γ tuned accordingly. The derived estimates for Y_u, Y_r and Y_{δ_r} are shown in Figs. 4.12, 4.13 and 4.14 respectively.

Aircraft system identification through flight testing for the UC^2AV [7] has shown that $Y_u \approx -288$. In this example, an initial condition of -100 is given for Y_u and

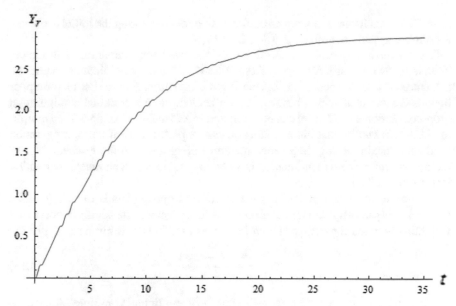

Fig. 4.13 Evolution of Y_r over time

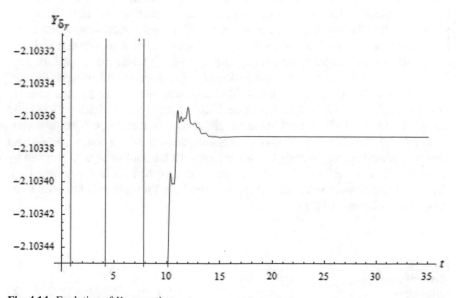

Fig. 4.14 Evolution of Y_{δ_r} over time

the system adapts to the true value of the unknown parameter (Fig. 4.12) via an appropriate tuning of the elements of matrix Γ, proving that the methodology is reliable. The benefit from this technique is twofold. It can be generalized and applied for estimation of the longitudinal/lateral dimensional control and stability derivatives of the UAV and it can also serve as an alternative for calculating the initial estimates for aircraft system identification using CIFER [9].

The methodology outlined above can be followed for estimation of the lower bounds (conventional UAV case) of Eq. (4.8) with the only requirement being the existence of a UAV model for XPlane flight simulation. Estimation of the upper bounds is more challenging because it requires that the upper bound configuration (non-conventional UAV case) can be simulated in XPlane. For example, for a morphing UAV this implies that flight simulation can be performed with the configuration producing maximum lift, drag, sideforce and rolling and heading moment. Nevertheless, aircraft system identification is a challenging task and parameter derivation is not 100% reliable.

Therefore, the effect of pushing the lower and upper bounds to safe (smaller and larger respectively) numerical values was investigated. The additive uncertainty weighting function for airspeed control that was used in [10] is given in Eq. (4.13).

$$W_{V_\alpha} = \frac{0.1s + 0.32}{s + 1.5} \tag{4.13}$$

This function was derived by considering the upper bound non-dimensional control and stability derivatives to be $C_{L_{0_2}} = 0.55$, $C_{D_{0_2}} = 0.05$, $C_{L_{q_2}} = 1.89$, $C_{L_{\delta_{e_2}}} = 0.09$. Controller performance in an extreme case scenario where these values are assumed to be 10 times larger was investigated. The results without and with tuned performance weighting function can be seen in Figs. 4.15 and 4.16, where it is clearly shown that the controller order remains the same (=3) while robust performance (peak-μ value < 1) can be guaranteed with an appropriate modification of the performance weighting function utilized. The argument used here is not a general, mathematically concrete and proven method but in a practical application it can yield acceptable reference signal tracking as the simulation results chapter demonstrates.

The section that follows presents the complete inner-outer loop, dynamic inversion plus μ-synthesis controller design methodology with the control objectives given by Eq. (4.14). The subscript ref represents reference instructions in the airspeed V_α, flight path angle γ and roll angle ϕ respectively. Sideslip angle β is intended to be close to 0 for lateral stability.

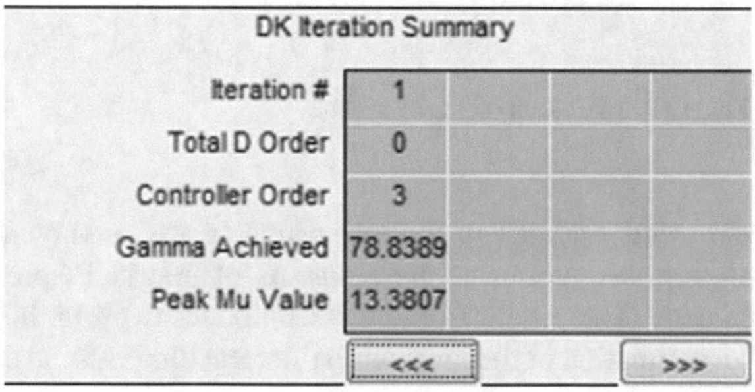

Fig. 4.15 Airspeed controller without tuned performance

DK Iteration Summary				
Iteration #	1			
Total D Order	0			
Controller Order	3			
Gamma Achieved	78.8389			
Peak Mu Value	13.3807			
<<<			>>>	

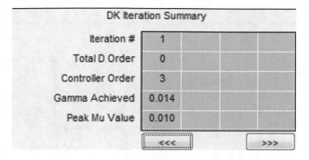

DK Iteration Summary				
Iteration #	1			
Total D Order	0			
Controller Order	3			
Gamma Achieved	0.014			
Peak Mu Value	0.010			

Fig. 4.16 Airspeed controller with tuned performance

$$V_\alpha = V_{\alpha_{ref}}$$
$$\gamma = \gamma_{ref}$$
$$\beta = 0 \qquad\qquad (4.14)$$
$$\phi = \phi_{ref}$$

4.3 Longitudinal Motion

The controller design process for the UAV longitudinal flight dynamics follows a cascaded structure, which is depicted in Fig. 4.17. First, a controller for V_α is designed using the engine thrust F_T as a control input and then the output is fed into the flight path angle controller, which is regulated through δ_e. State variables q and α are assumed to be available for measurement throughout the controller synthesis and $V_{\alpha_{ref}}$ and γ_{ref} represent reference instructions for V_α and γ respectively.

4.3.1 Airspeed Controller

An inner-loop dynamic inversion controller is employed to partially linearize the airspeed dynamics. After that, the plant is separated into unknown (uncertainty) and known parts and D-K iteration takes place to control the overall uncertain system. The way the airspeed controller operates can be seen in Fig. 4.18.

The airspeed dynamics from [2] read:

$$\dot{V}_\alpha = \frac{1}{m}(-F_{drag} + F_T \cos\alpha - mg \sin\gamma) \qquad (4.15)$$

with the mathematical expression for the drag force given by:

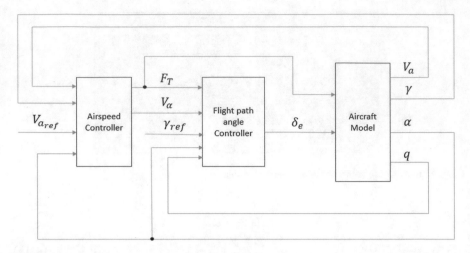

Fig. 4.17 Block diagram for longitudinal motion control

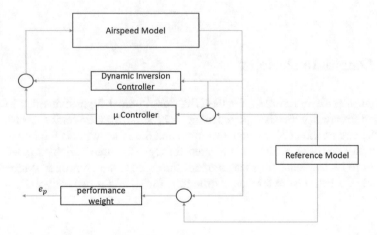

Fig. 4.18 Airspeed controller design framework

$$F_{drag} = \frac{1}{2}\rho V_\alpha^2 S C_D \qquad (4.16)$$

Using (4.6), (4.8) and (4.16) we alter the airspeed dynamics in the following way:

$$\dot{V}_\alpha = -\frac{1}{2m}\rho V_\alpha^2 S \left(C_{D_{0_1}} + \frac{(C_{L_{0_1}} + C_{L_{\alpha_1}}\alpha)^2}{\pi e AR} \right)$$

$$+ \frac{1}{m}\cos\alpha \cdot F_T - g\sin\gamma + \sum_{i=1}^{2}\Delta_{f_i} \qquad (4.17)$$

Fig. 4.19 Uncertain open-loop interconnection P for the airspeed dynamics [4]

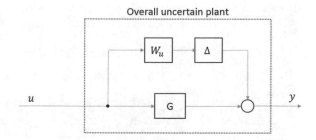

where Δ_{f_i}, $i = 1, 2$ are additive uncertainties of the functions

$$f_1 = -\frac{1}{2m}\rho V_\alpha^2 S \left(C_{D_{0_1}} + \frac{\left(C_{L_{0_1}} + C_{L_{\alpha_1}}\alpha \right)^2}{\pi e AR} \right)$$

$$f_2 = \frac{1}{m}\cos\alpha \cdot F_T \tag{4.18}$$

Functions f_1, f_2 are generated by the additional effect of aerodynamic uncertainty and the added load that is considered, respectively. First, the following dynamic inversion inner-loop control law is applied:

$$F_T = \frac{m}{\cos\alpha}(g \sin\gamma + f_1 + v_1) \tag{4.19}$$

with v_1 being a pseudo control input. The airspeed dynamics reduce to:

$$\dot{V}_\alpha = v_1 + \sum_{i=1}^{2} \Delta_{f_i} \tag{4.20}$$

Choosing $v_1 = k_{V_\alpha}(V_{\alpha_{ref}} - V_\alpha)$ with k_{V_α} being a design tuning parameter, yields an uncertain plant of the form:

$$\tilde{G}(s) = \frac{k_{V_\alpha}}{s + k_{V_\alpha}} + \frac{\mathscr{L}(\sum_{i=1}^{2}\Delta_{f_i})}{s + k_{V_\alpha}} \tag{4.21}$$

where $\mathscr{L}$ is the Laplace transform symbol. The airspeed's nominal model is given by: $G(s) = \frac{k_{V_\alpha}}{s+k_{V_\alpha}}$. The uncertain open-loop interconnection P for the airspeed dynamics is depicted in Fig. 4.19.

Where the perturbation block Δ must satisfy the norm condition $\|\Delta\| \le 1$ and the additive uncertainty weighting function W_u is determined by (4.22) and by making use of (4.8) and (4.9).

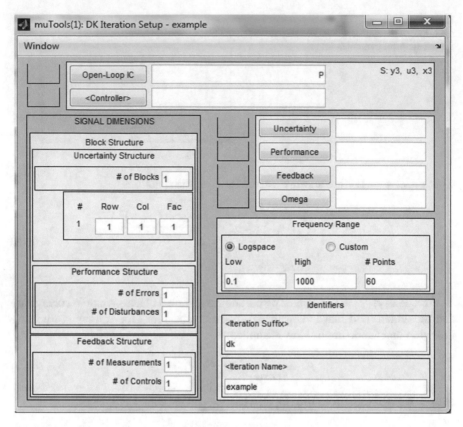

Fig. 4.20 μ-Analysis and Synthesis Toolbox interface [4]

$$\left|\tilde{G}(j\omega) - G(j\omega)\right| \leq |W_u| \tag{4.22}$$

Having derived the nominal model $G(s)$, the additive uncertainty weighting function $W_u(s)$ and the open loop interconnection P for the UAV airspeed, the outer-loop μ-controller design process can be initiated using the μ-Analysis and Synthesis Toolbox [4] by running the D-K iteration graphic user interface (GUI) to derive the μ-synthesis control law. For newer MATLAB versions, the plots and the design process can be recreated using the Robust Control Toolbox [11] for μ-synthesis design. Figure 4.20 illustrates how the graphic user interface looks and how it functions.

In order to apply the general structured singular value theory to control system design, the control problem needs to be recast into the linear fractional transformation (LFT) setting as in Fig. 4.21.

Fig. 4.21 Linear fractional transformation setting for μ-synthesis design [4]

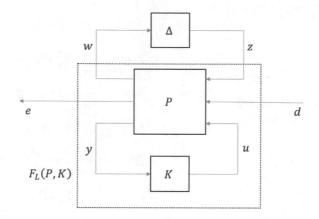

$F_L(P, K)$

The μ-controller K is derived through D-K iteration, the process of which is represented in (4.23).

$$\min_{K \ stabilizing} \ \max_{\omega} \ \min_{D_\omega \in D_\Delta} \ \bar{\sigma}[D_\omega F_L(P, K)(j\omega)D_\omega^{-1}] \qquad (4.23)$$

where $\bar{\sigma}$ denotes the maximum singular value and D_Δ is the set of matrices that satisfy $D_\Delta \cdot \Lambda = \Lambda \cdot D_\Delta$.

4.3.2 Flight Path Angle Controller

The flight path angle controller follows an identical approach with the airspeed controller. The flight path angle dynamics from [2] read:

$$\dot{\gamma} = \frac{1}{mV_\alpha}(F_{lift} + F_T \sin \alpha - mg \cos \gamma) \qquad (4.24)$$

with the relation between the lift force F_{lift} and the respective non-dimensional lift coefficient C_L, as found in [1], given by:

$$F_{lift} = \frac{1}{2}\rho V_\alpha^2 S C_L \qquad (4.25)$$

Using (4.6), (4.8) and (4.25) and defining the functions $f_i, i = 3, 4, \ldots, 8$ as in (4.26)

$$f_3 = \frac{1}{2mV_\alpha} \rho V_\alpha^2 S C_{L_{0_1}}$$

$$f_4 = \frac{1}{2mV_\alpha} \rho V_\alpha^2 S C_{L_{\alpha_1}} \alpha$$

$$f_5 = \frac{1}{2mV_\alpha} \rho V_\alpha^2 S C_{L_{q_1}} q$$

$$f_6 = \frac{1}{mV_\alpha} F_T \sin \alpha \qquad (4.26)$$

$$f_7 = \frac{g \cos \gamma}{V_\alpha}$$

$$f_8 = \frac{1}{2mV_\alpha} \rho V_\alpha^2 S C_{L_{\delta e_1}}$$

we can represent the flight path dynamics as in (4.27).

$$\dot{\gamma} = \sum_{i=3}^{6} f_i - f_7 + f_8 \cdot \delta_e + \sum_{i=3}^{6} \Delta_{f_i} + \Delta_{f_8} \delta_e \qquad (4.27)$$

Applying the dynamic inversion law $\delta_e = \frac{1}{f_8}(-\sum_{i=3}^{6} f_i + f_7 + v_2)$ and choosing $v_2 = k_\gamma(\gamma_{ref} - \gamma)$ transforms the flight path dynamics into a system appropriate for μ-synthesis design. The μ-Analysis and Synthesis Toolbox is then used to derive the outer-loop μ-synthesis flight path angle controller.

4.4 Lateral Motion

The same design is followed for sideslip angle control with the sideslip nonlinear dynamics given by Eq. (4.28) and taken from [1]. Rudder deflection control input δ_r is used to regulate sideslip.

$$\dot{\beta} = pw - ru + g \cos \theta \sin \phi + \frac{\rho V_\alpha^2 S}{2m} C_Y \qquad (4.28)$$

The roll controller has a fundamental difference in the order of the nominal plant and the order of the designed controller. The roll UAV dynamics are mathematically described by a first order, nonlinear differential equation that does not contain the aileron as control input as shown in Eq. (4.29).

$$\dot{\phi} = p + \tan \theta (q \sin \phi + r \cos \phi) \qquad (4.29)$$

This is tackled mathematically by taking the derivative of the roll dynamics and taking into account the roll rate p dynamics as shown in Eq. (4.30) and the mathematical expression of C_p in Eq. (4.6).

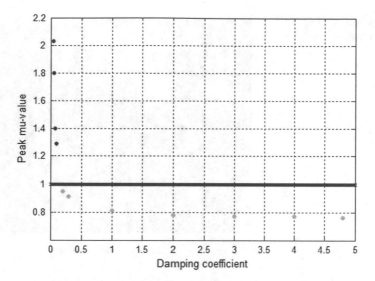

Fig. 4.22 Damping coefficient versus peak μ-value for roll nominal plant

$$\dot{p} = \Gamma_1 pq - \Gamma_2 qr + \frac{1}{2}\rho V_\alpha^2 SbC_p \qquad (4.30)$$

The aircraft aileron can be used as the control input to apply the inner-loop dynamic inversion controller as before. Accurate control of the roll angle of the UAV is sought and the roll dynamics nominal plant in this case will be a standard second order transfer function. The choice of the type (underdamped, overdamped, etc.) of the roll nominal plant is not straightforward as before, where it was defined by the dynamic-inversion design tracking parameter (i.e., k_{V_α}). Choosing a natural frequency of $\omega_\phi = 1$ rad/s, the effect of the damping coefficient ζ on the roll nominal plant D-K optimization was investigated and it is depicted in Fig. 4.22.

The plot is a graphical representation of the damping coefficient versus the peak μ-value achieved after the optimization process, with the critical line of $y = 1$ highlighted in blue. The system's peak μ-value drops in exponential fashion until it settles to the optimal value of $\zeta_\phi = 4.8$ (overdamped). The red markers correspond to rejected and the green markers to accepted numerical values based on the criterion of peak μ-value < 1.

4.5 SIL Environment and Sensor Models

X-Plane models of the RMRC Anaconda and UC^2AV (Fig. 4.23) are created using JavaFoil, Airfoil Maker, and Plane Maker (Fig. 4.24). JavaFoil is used to model the airfoil aerodynamic properties and dimensions at the appropriate Reynold's number. UC^2AV design details and airfoil properties can be found in [12]. Segmenting the

Fig. 4.23 The UC^2AV model in X-Plane [7]

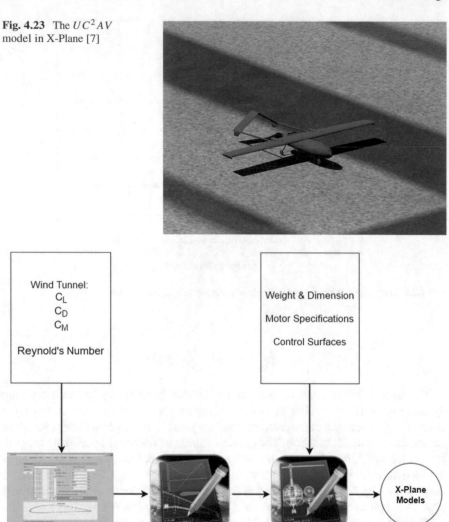

Fig. 4.24 Workflow for creating UAV X-Plane models. The coefficients of lift C_L, drag C_D, and moment C_m are modeled by X-Plane [7]

RMRC Anaconda wing and measuring the airfoil dimensions (maximum thickness location) revealed that the RMRC Anaconda has a Clark YH airfoil (Fig. 4.25).

The modeled airfoils are converted to X-Plane using X-Plane's Airfoil Maker. Plane Maker is used to model the weight and dimensions of the UAV parts (fuselage, wings, landing gear, etc.), the location of center of gravity, motor specifications (horse power, maximum rpm, etc.), and control surface dimensions (control surface width, maximum deflection angles, location on wings, etc.).

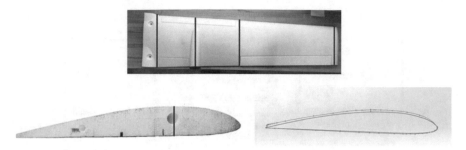

Fig. 4.25 Top: segmented wing for airfoil property analysis. Bottom left: location of maximum airfoil thickness. Bottom right: Clark YH airfoil [7]

Fig. 4.26 A UDP packet sent by X-Plane to Simulink [7]

Simulink code is written to communicate with X-Plane via the UDP protocol. The data generated by X-Plane (control inputs and UAV response) is packeted and sent to Simulink. The packeted data is in decimal code and single precision floating point format (Fig. 4.26). Bytes 1–4 contain the Header, byte 5 is a software internal use byte, bytes 6–10 contain the data label and bytes 11–41 contain the estimated aerodynamic variables. The Simulink code unpacks and converts the data to decimal values (Fig. 4.27) [13]. The values are also plotted in real-time to provide immediate feedback (Fig. 4.28).

It is important to note that Simulink's simulation time is dependent on the complexity of the code and the speed of computer. The time array produced by Simulink can be inaccurate leading to incorrect data time stamping. To resolve the issue, tic and toc MATLAB functions are used. Input signals are applied to the aircraft using the Interlink Elite Controller and UAV response to the inputs is recorded using Simulink and MATLAB.

The following section provides the basic steps, the framework and the necessary technical preliminaries for Kalman, linear and nonlinear complementary filter design in Simulink. The section is divided into three subsections. One for the noise models that are utilized and one section for Kalman and complementary filter implementation respectively.

4.5.1 Noise Models

Throughout the design, pitch, roll and heading (yaw) readings that are received by Simulink from X-Plane (Fig. 4.27) are corrupted by noise before being sent to the observer framework as inputs. Noise is by default generated to reflect a 10% sensor inaccuracy based on the estimated attitude angle. MATLAB functions $wgn()$ and

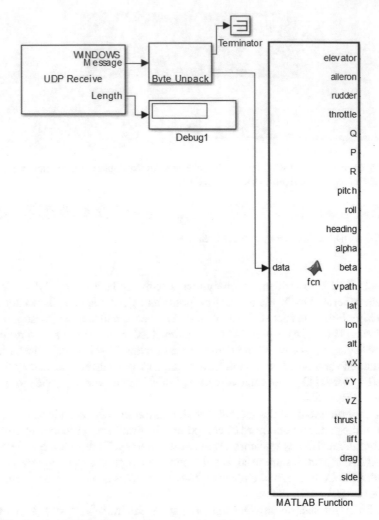

Fig. 4.27 Simulink UDP receiver and data conversion [7]

$rand()$ are used to generate white, Gaussian and uniform random noise respectively. A normalized preview of the noise models used in the design can be seen in Figs. 4.29 and 4.30.

4.5.2   Kalman Filter

The Kalman filter framework for pitch angle estimation in Simulink is summarized by Fig. 4.31. This is essentially a Simulink subsystem (5 inputs, 1 output) designed

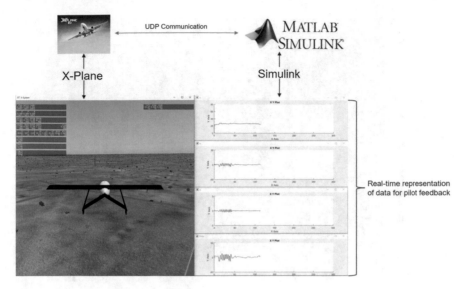

Fig. 4.28 Real-time visualization of the recorded data [7]

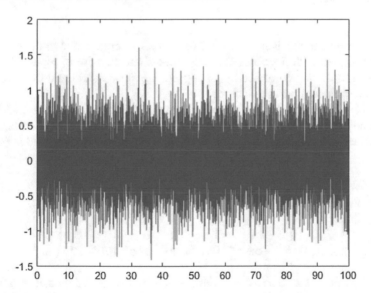

Fig. 4.29 White Gaussian noise model

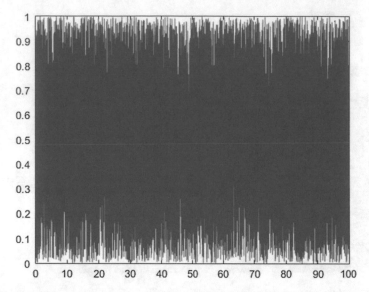

Fig. 4.30 Uniform noise model

on top of the UDP receiver shown in Fig. 4.27. A discrete Kalman filter Simulink function is used with 5 system states, 5 outputs and no control input. The actual subsystem inputs are corrupted by an appropriate noise model and then fed to the Kalman filter input port as model source.

Matrix A for state-space model representation is derived from the UAV longitudinal and lateral state-space models that can be found in [1]. For instance, the longitudinal state-space model corresponding to Fig. 4.31 is highlighted in (4.31).

$$\begin{pmatrix} \dot{u} \\ \dot{w} \\ \dot{q} \\ \dot{\theta} \\ \dot{h} \end{pmatrix} = \begin{pmatrix} X_u & X_w & X_q & -g\cos\theta^* & 0 \\ Z_u & Z_w & Z_q & -g\sin\theta^* & 0 \\ M_u & M_w & M_q & 0 & 0 \\ 0 & 0 & 1 & 0 & 0 \\ \sin\theta^* & -\cos\theta^* & 0 & u^*\cos\theta^* + w^*\sin\theta^* & 0 \end{pmatrix} \begin{pmatrix} u \\ w \\ q \\ \theta \\ h \end{pmatrix} \tag{4.31}$$

Matrix C is chosen to be a 5×5 identity matrix. Nomenclature and details about the aerodynamic parameters appearing as matrix elements in (4.31) can be found in [1]. The non-dimensional aerodynamic coefficients that are related to matrix A are calculated in [10]. Roll and heading angle estimation via the Kalman method in this SIL environment follows the same concept by utilizing the lateral state-space model of the UAV.

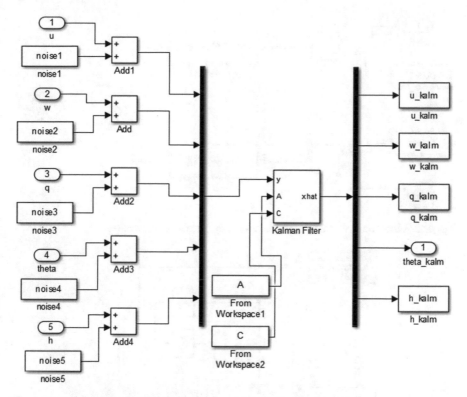

Fig. 4.31 Kalman filter framework for pitch angle estimation

4.5.3 Complementary Filter

The linear complementary filter framework for pitch angle estimation in Simulink is illustrated by Fig. 4.32, implementing a simple low-pass, high-pass filter structure. A low-pass filter $\frac{1}{\tau s+1}$ compensates for low frequencies for pitch rate measurement from a gyro while a high-pass filter $\frac{\tau s}{\tau s+1}$ takes care of the high frequencies for pitch angle from an accelerometer. The time constant τ that determines the cut-off frequency is derived experimentally, after running a few tests and evaluating the filter performance. Roll and heading angle estimation is performed by following the same design.

However, the linear filter does not produce reliable estimates when the system under consideration is nonlinear and the sensor has varying bias. The nonlinear complementary filter is thus designed to take cues from the orientation error between complementary sensor modules and improve the estimation. The block diagram for estimating attitude using nonlinear complementary filters is shown above in Fig. 4.33. Implementation in Simulink for pitch angle estimation is illustrated in Fig. 4.34. Proportional and integral constants K_p, K_I are again determined experimentally based on estimation efficiency.

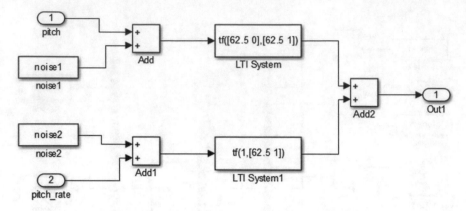

Fig. 4.32   Linear complementary filter framework for pitch angle estimation

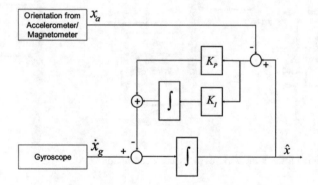

Fig. 4.33   Nonlinear complementary filter structure [14]

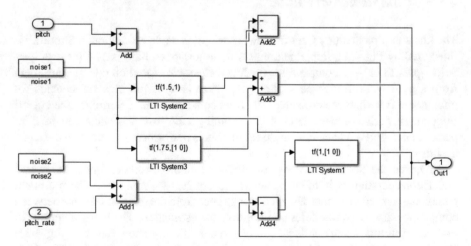

Fig. 4.34   Nonlinear complementary filter framework for pitch angle estimation

Table 4.1 Constants used for complementary filter design

	K_P	K_I	τ	f_c (Hz)
Pitch	1.5	1.75	62.5	0.0024
Roll	6.21	3.84	57	0.0031
Heading	2.92	4.3	67.1	0.0015

Table 4.1 shows the proportional, integral, time and cut-off frequency (f_c) constants for pitch, roll and heading nonlinear complementary filter design.

4.6 Controller Implementation in Simulink

Details, block diagrams and an in-depth description of how the inner-outer loop dynamic inversion plus μ-synthesis controller is implemented in Simulink are included in this section.

The controller design methodology for airspeed control is outlined step-by-step, with flight path, sideslip and roll angle control following the same steps. The μ-Analysis and Synthesis Toolbox produces the outer-loop μ-synthesis airspeed controller in the frequency domain. The modeling functions shown in Fig. 4.35 are used to convert the controller generated by the toolbox to easily accessible state-space representation that is used to create the actual controller block in Simulink, illustrated in Fig. 4.36. Specifically, the functions $sys2pss$ and $unpck$ are used for the state-space conversion process. The outer-loop controller is basically a linear, time-invariant system with a transfer function that corresponds to the controller methodology presented in Sect. 4.3.1. It contains two inputs, airspeed and airspeed reference instruction and

Modeling Functions	
mfilter	Construct a Bessel, Butterworth, Chebychev, or RC filter
nd2sys	Convert a SISO transfer function into a μ-Tools **S** matrix
pck	Create a **S** matrix from state-space data (A, B, C, D)
pss2sys	Convert an [A B;C D] matrix into a μ-Tools **S** matrix
sys2pss	Extract state-space matrix [A B; C D] from a **S** matrix
sysic	System interconnection program
unpck	Extract state-space data (A,B,C,D) from a **S** matrix
zp2sys	Convert transfer function poles and zeros to a **S** matrix

Fig. 4.35 Modeling functions for extracting state-space data [4]

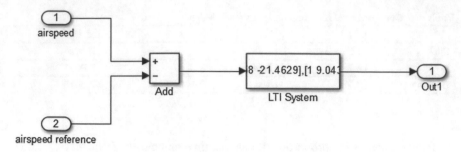

Fig. 4.36 Outer-loop μ-synthesis airspeed controller in Simulink

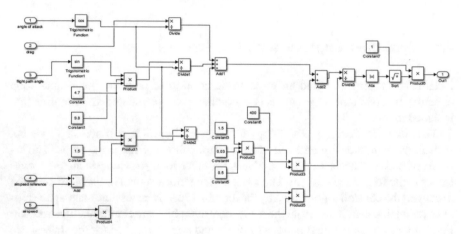

Fig. 4.37 Inner-loop dynamic inversion airspeed controller in Simulink

one output, the newly generated airspeed that will be fed into the dynamic inversion, inner-loop controller.

The dynamic-inversion, inner-loop, airspeed controller implementation is illustrated in Fig. 4.37. The blocks are implementing the math required to linearize the airspeed dynamics as described in Sect. 4.3.1. Putting the two controllers together, forms the proposed, general and universally applicable control architecture that is highlighted in Fig. 4.38.

Figure 4.39 shows how Sects. 4.3, 4.4 and 4.5 are integrated together. In other words, how wind models and atmospheric disturbance components are entering the UAV equations of motion in Simulink. Parameters w_{n_s}, w_{e_s}, w_{d_s} correspond to wind in NED directions, subscript w_g refers to parameters with respect to the wind frame and R_v^b is the rotation matrix from the vehicle to the body frame.

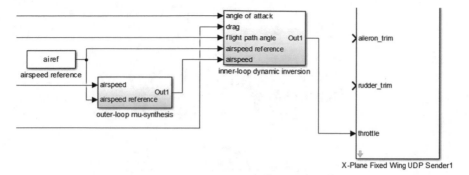

Fig. 4.38 Overview of the airspeed controller in Simulink

Fig. 4.39 Wind models and atmospheric disturbances in Simulink [1]

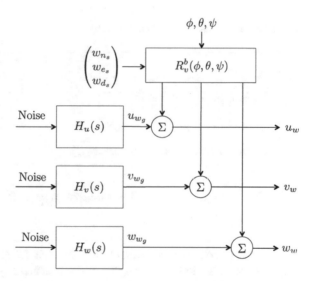

Filters H_u, H_v, H_w are Dryden transfer functions for approximating turbulence models, with the exact mathematical definition given by Fig. 4.40. Parameters σ_u, σ_v, σ_w are the intensities of turbulence and L_u, L_v, L_w represent spatial wavelengths.

Pitch, roll and heading angle readings are corrupted with wind, passed through the Kalman or complementary filter sensor model, blended with Dryden turbulence models and then generating the appropriate linear velocities that are used for controller design in Sects. 4.3 and 4.4. Figure 4.41 contains all the numerical values for each gust model case and focus is placed particularly on the medium altitude, moderate turbulence, for small-scale fixed-wing UAVs.

$$H_u(s) = \sigma_u \sqrt{\frac{2V_a}{L_u}} \frac{1}{s + \frac{V_a}{L_u}}$$

$$H_v(s) = \sigma_v \sqrt{\frac{3V_a}{L_v}} \frac{\left(s + \frac{V_a}{\sqrt{3}L_v}\right)}{\left(s + \frac{V_a}{L_v}\right)^2}$$

$$H_w(s) = \sigma_w \sqrt{\frac{3V_a}{L_w}} \frac{\left(s + \frac{V_a}{\sqrt{3}L_w}\right)}{\left(s + \frac{V_a}{L_w}\right)^2}$$

Fig. 4.40 Dryden gust transfer functions [1]

gust description	altitude (m)	$L_u = L_v$ (m)	L_w (m)	$\sigma_u = \sigma_v$ (m/s)	σ_w (m/s)
low altitude, light turbulence	50	200	50	1.06	0.7
low altitude, moderate turbulence	50	200	50	2.12	1.4
medium altitude, light turbulence	600	533	533	1.5	1.5
medium altitude, moderate turbulence	600	533	533	3.0	3.0

Fig. 4.41 Dryden transfer function parameters [1]

4.7 Remarks—Controller Design Algorithm—Stability Analysis

This section provides an overview of the controller design algorithm with the general steps shown in Table 4.2. Starting from the nonlinear system dynamics (Step 1), a nominal (known) plus uncertain (unknown) part representation is derived based on the uncertainty interval and the lower bounds for the aerodynamic coefficients (Step 2). The inner-loop, dynamic inversion controller is then designed based on the expanded dynamics of Step 2 (Step 3), partially linearizing the system. With a reduced linear system form available, the dynamics are modeled using an overall uncertain plant ($\tilde{G}$) and a nominal plant (G), with the absolute value of the difference of these two yielding the additive uncertainty weighting function for the respective aerodynamic uncertainty intervals (Steps 4, 5). The final step is the design of the outer-loop μ-synthesis controller which is performed by a doubly iterative procedure with respect to (i) the H_∞ controller K and (ii) the left and right D-scales to optimize the system's (transfer matrix) maximum singular value $\bar{\sigma}$ (Step 6).

Steps 1 through 6 are followed for nonlinear control of airspeed, flight path angle, sideslip and roll angle. For stability analysis purposes, Steps 3, 4 and 6 require attention. Specifically, the nonlinear dynamics are linearized and a pseudo control input is inserted into the design, allowing one to incorporate the tracking error into the simplified, linear dynamics. Parameter v is a function of the tracking error.

Table 4.2 Controller design algorithm

Steps	Task description
Step 1	Nonlinear dynamics: $\dot{x} = f(x) + g(x)u$
Step 2	Step 1 into nominal/uncertain part: $\dot{x} = \hat{f}(x) + \Delta f(x) + [\hat{g}(x) + \Delta g(x)]u$
Step 3	Dynamic inversion controller to step 2: $u^* = \hat{g}(x)^{-1}[v - \hat{f}(x)]$
Step 4	Reduced system dynamics: $\dot{x} = v + \Delta f(x) + \Delta g(x)u^*$
Step 5	Additive uncertainty weighting functions: $\left\lvert \tilde{G}(j\omega) - G(j\omega) \right\rvert \leq \lvert W_u \rvert$
Step 6	μ-synthesis optimization: $\displaystyle \min_{K \ stabilizing} \ \max_{\omega} \ \min_{D_\omega \in D_\Delta} \bar{\sigma}[D_\omega F_L(P, K)(j\omega)D_\omega^{-1}]$

The μ-synthesis optimization process is there to guard against instabilities that may occur for when the reduced system dynamics are unstable. So in principle, stability is investigated for a linear system with time-varying uncertainties (functions $\Delta f(x)$, $\Delta g(x)$) or perturbations.

References

1. Beard RW, McLain TW (2012) Small unmanned aircraft: theory and practice. Princeton University Press, Princeton
2. Stevens BL, Lewis FL, Johnson EN (2015) Aircraft control and simulation: dynamics, controls design, and autonomous systems. Wiley, New York
3. Lesprier J, Biannic JM, Roos C (2015) Modeling and robust nonlinear control of a fixed-wing UAV. In: IEEE conference on control applications (CCA), pp 1334–1339
4. Balas GJ, Doyle JC, Glover K, Packard A, Smith R (1993) μ-analysis and synthesis toolbox. MUSYN Inc. and The MathWorks
5. Klein V, Morelli EA (2006) Aircraft system identification: theory and practice. American Institute of Aeronautics and Astronautics, Reston
6. Ioannou PA, Sun J (2010) Robust adaptive control. Prentice-Hall, Upper Saddle River (Reprint)
7. Agha M (2017) System identification of a circulation control unmanned aerial vehicle. Doctoral dissertation, University of Denver
8. Zhi J, Dong X, Chen Y, Liu Z, Shi C (2017) Robust adaptive finite time parameter estimation with relaxed persistence of excitation. In: IEEE Asian control conference (ASCC), pp 1384–1388
9. Tischler MB, Remple RK (2006) Aircraft and rotorcraft system identification. AIAA education series. American Institute of Aeronautics and Astronautics, Reston
10. Michailidis MG, Kanistras K, Agha M, Rutherford MJ, Valavanis KP (2018) Nonlinear control of fixed-wing UAVs with time-varying aerodynamic uncertainties via μ-synthesis. In: IEEE conference on decision and control (CDC), pp 6314–6321
11. Balas G, Chiang R, Packard A, Safonov M (2005) Robust control toolbox, for use with Matlab, User's Guide
12. Kanistras K, Rutherford MJ, Valavanis KP (2018) Foundations of circulation control based small-scale unmanned aircraft. Springer, Berlin
13. Bittar A, Figuereido HV, Guimaraes PA, Mendes AC (2014) Guidance software-in-the-loop simulation using x-plane and simulink for UAVs. In: IEEE international conference on unmanned aircraft systems (ICUAS), pp 993–1002
14. Kottath R, Narkhede P, Kumar V, Karar V, Poddar S (2017) Multiple model adaptive complementary filter for attitude estimation. Aerosp Sci Technol 69:574–581

Chapter 5
UC^2AV Case Study

Abstract The proposed controller design framework has been validated for stability and navigational control of the UC^2AV. For CC to be implemented on the UC^2AV, a forward impeller centrifugal compressor is used, located in the fuselage and called Air Supply Unit (ASU), while an Air Delivery System (ADS) integrated with a plenum is capable of distributing air uniformly across the wingspan [1]. CC is applied through the ASU by regulating the RPM of the centrifugal compressor. For future missions and scenarios, the RPM of the ASU will need to be optimally controlled (CC-on-demand) with respect to power consumption or mission performance, to complete a variety of tasks. As a result, the RPM of the centrifugal compressor ranges between 0 and maximum (28,000) according to the ongoing mission. Different RPM will generate different values for the aerodynamic coefficients of the UC^2AV, which will, in turn, generate different UC^2AV flight dynamics. Therefore, the actual UC^2AV flight dynamics are explicitly described by a family of models or by model uncertainty mainly stemming from the aerodynamic coefficients, with predefined upper and lower bounds.

5.1 Effects of CC on the UC^2AV Model

The impact of the implementation of CC on the UC^2AV flight dynamics is what distinguishes this work from existing studies conducted on navigation and control of UAVs. A UC^2AV based controller design that takes into account the effects of CC is investigated. Hence, this section presents the changing parameters due to CC in the UC^2AV model.

5.1.1 Mass

The mass of the airframe changes due to the existence of the ASU (Air Supply Unit) and the ADS (Air Delivery System) that are both located in the fuselage of the aircraft. With this type of equipment on-board, the UC^2AV weighs 4.7 kg, (i.e., 1.2 kg of

© Springer Nature Switzerland AG 2020

M. G. Michailidis et al., *Nonlinear Control of Fixed-Wing UAVs*
with Time-Varying and Unstructured Uncertainties, Springer Tracts
in Autonomous Systems 1, https://doi.org/10.1007/978-3-030-40716-2_5

additional weight compared to the stock RMRC Anaconda). Since CC enhances the payload capabilities of the platform, for the aircraft's mass m, a package delivering scenario is a promising and realistic application.

5.1.2   Moments and Products of Inertia

According to the fixed-wing UAV equations of motion found in [2], the UC^2AV flight dynamics are affected by parameters Γ_i, which represent functions of the aircraft's moments and products of inertia. Furthermore, the moments and products of inertia depend on the aircraft's total mass. Due to the modified mass distribution in the fuselage, caused by the ASU and the ADS, the total mass of the airframe changes as well. This results in a change of the moments and products of inertia of the UC^2AV.

5.1.3   Lift Coefficient

The relation between the lift force F_{lift} and the respective non-dimensional lift coefficient C_L as found in [2] is:

$$F_{lift} = \frac{1}{2}\rho V_\alpha^2 S C_L \tag{5.1}$$

where ρ is the air density, V_α is the UC^2AV's airspeed and S is the surface area of the wing. CC is a method to enhance lift and this can be seen as an increase on the lift coefficient of the airfoil C_L. So graphically, the effect of CC can be interpreted as a translation of the lift coefficient versus angle of attack graph. This is shown in Fig. 5.1, which illustrates a 2D wind tunnel experimental study of the effect of CC on C_l based on different values of the momentum coefficient of blowing C_μ [3].

5.1.4   Drag Coefficient

The mathematical expression for the drag force is given by:

$$F_{drag} = \frac{1}{2}\rho V_\alpha^2 S C_D \tag{5.2}$$

where C_D denotes the drag coefficient of the UC^2AV. The total drag of the aircraft is a summation of the parasitic drag, the induced drag and the wave drag. The induced drag, also called vortex drag, is the pressure drag caused by the tip vortices of a finite wing when it is producing lift [4]. CC alters the induced drag of the aircraft and this is interpreted as a change on the overall drag coefficient, C_D. The exact expression

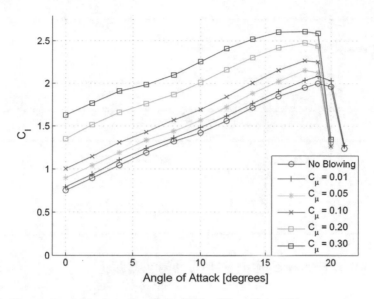

Fig. 5.1 2D experimental study on the effect of CC on lift coefficient [3]

of the induced drag is:

$$C_{D_i} = \frac{C_L{}^2}{\pi eAR} \qquad (5.3)$$

where e is the Oswald efficiency factor and the wing aspect ratio, taken from [4], is given by (5.4).

$$AR = \frac{b^2}{S} \qquad (5.4)$$

5.1.5 Pitching Moment Coefficient

The exact expression for the pitching moment coefficient can be found in [4] but it is restated here for completeness:

$$C_m = f(\alpha, \delta_e, q) + \frac{x_R}{\bar{c}} C_L \qquad (5.5)$$

Equation (5.5) highlights the dependence of the pitching moment coefficient on the lift coefficient. Since CC changes the lift coefficient, it also affects the dynamic behavior of the pitching moment coefficient, causing an additional pitching moment effect and a dramatic change on the aircraft's pitch angle. The effect has been recently validated [3], through flight testing and data collection. It can be seen in Fig. 5.2, with the red dashed lines indicating the intervals corresponding to the effect.

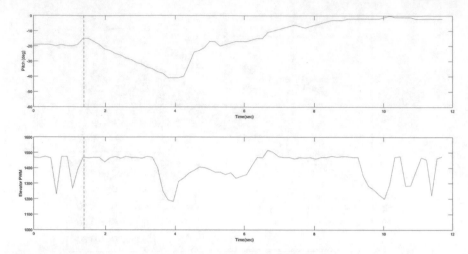

Fig. 5.2 Change of pitch angle due to CC during cruise flight [3]

5.2 Kalman and Complementary Filter Attitude Estimation

This section presents a software-in-the-loop filter performance results under Gaussian and uniform noise, implemented for attitude estimation during cruise, circular and an aggressive slalom maneuver. The maneuvers were performed using the Interlink Elite Controller. Figures 5.3, 5.4, 5.5, 5.6, 5.7 and 5.8 correspond to cruise flight, Figs. 5.9, 5.10, 5.11, 5.12, 5.13 and 5.14 depict circular maneuver results and finally, Figs. 5.15, 5.16, 5.17, 5.18, 5.19 and 5.20 correspond to the slalom maneuver. Overall, the Kalman filter provides acceptable accuracy but fails to track challenging nonlinearities. In this regard, both linear and nonlinear complementary filters provide superior performance. However, a significant difference between the two is observed when it comes to attitude estimation error reduction, with the nonlinear framework achieving an almost ideal performance. Finally, the results show that the type of noise can always be tackled by an appropriate optimization of the design, both for Kalman and complementary filters.

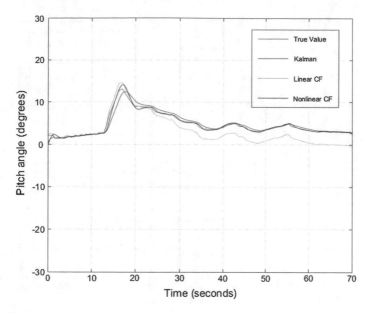

Fig. 5.3 Pitch angle estimation under Gaussian noise for cruise

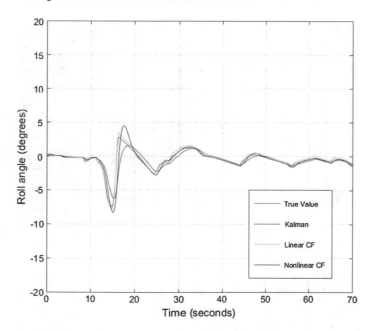

Fig. 5.4 Roll angle estimation under Gaussian noise for cruise

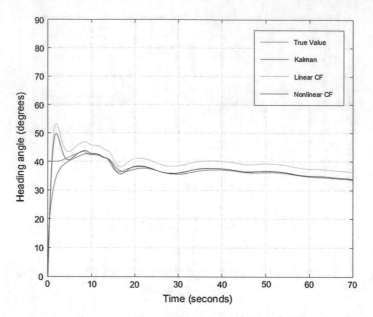

Fig. 5.5 Heading angle estimation under Gaussian noise for cruise

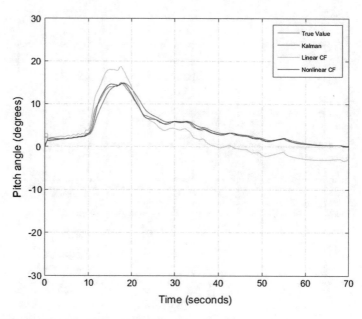

Fig. 5.6 Pitch angle estimation under uniform noise for cruise

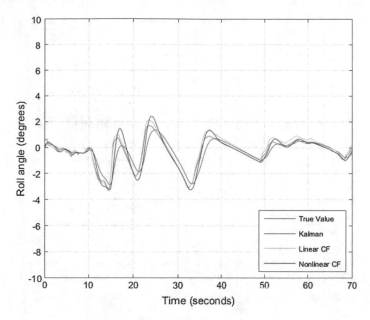

Fig. 5.7   Roll angle estimation under uniform noise for cruise

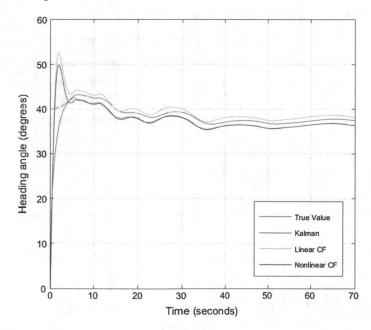

Fig. 5.8   Heading angle estimation under uniform noise for cruise

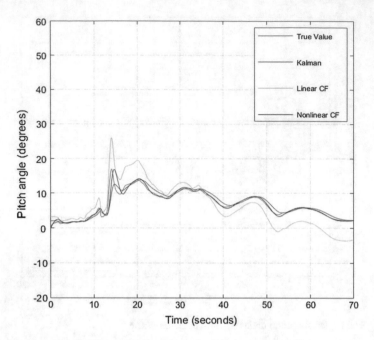

Fig. 5.9 Pitch angle estimation under Gaussian noise for circular

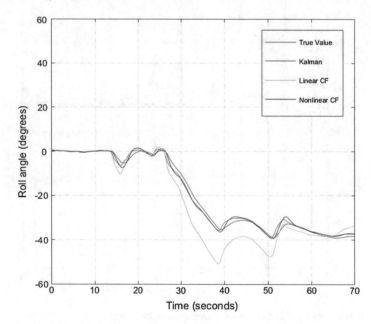

Fig. 5.10 Roll angle estimation under Gaussian noise for circular

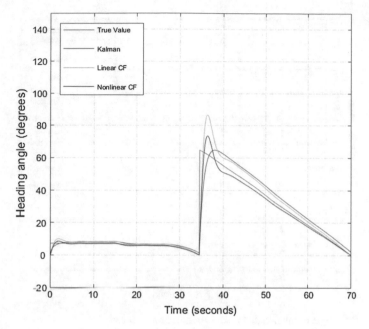

Fig. 5.11   Heading angle estimation under Gaussian noise for circular

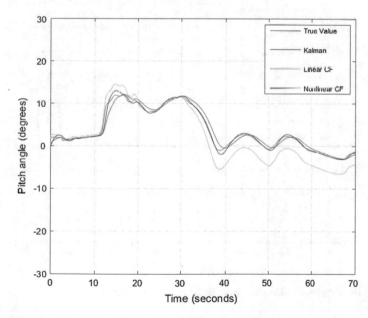

Fig. 5.12   Pitch angle estimation under uniform noise for circular

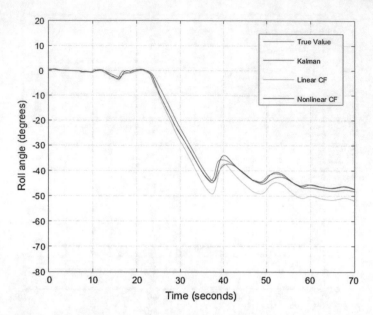

Fig. 5.13 Roll angle estimation under uniform noise for circular

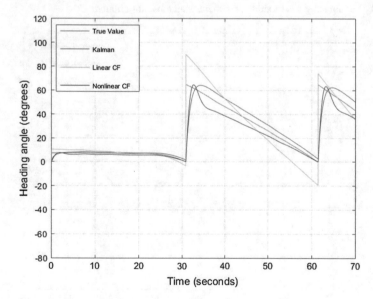

Fig. 5.14 Heading angle estimation under uniform noise for circular

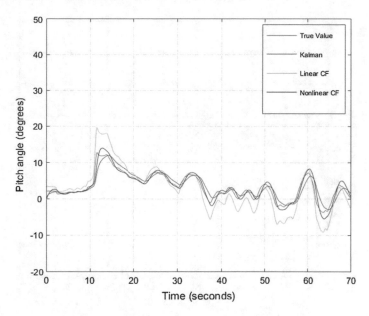

Fig. 5.15   Pitch angle estimation under Gaussian noise for slalom

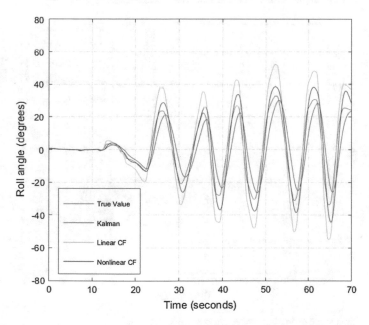

Fig. 5.16   Roll angle estimation under Gaussian noise for slalom

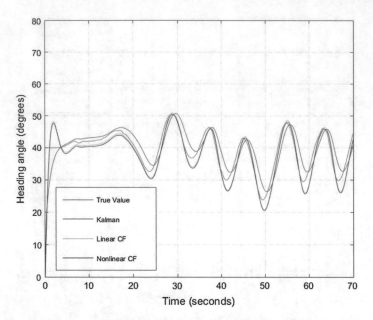

Fig. 5.17   Heading angle estimation under Gaussian noise for slalom

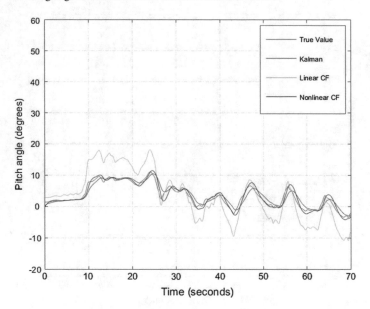

Fig. 5.18   Pitch angle estimation under uniform noise for slalom

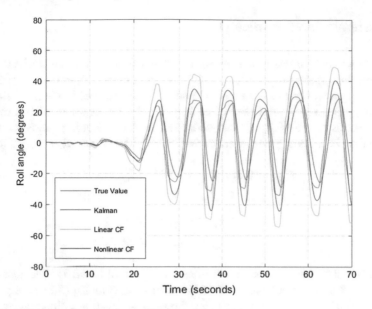

Fig. 5.19 Roll angle estimation under uniform noise for slalom

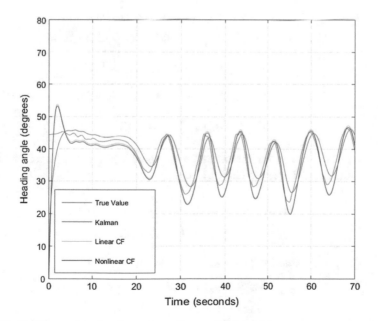

Fig. 5.20 Heading angle estimation under uniform noise for slalom

5.3   Simulation Parameters

The proposed design was implemented for nonlinear control of the UC^2AV [5, 6]. The inertia parameters $\Gamma_1 = 0.1, \Gamma_2 = 0.72, \Gamma_3 = 1.63$ and $\Gamma_4 = 0.09$ were derived by treating the aircraft as a sum of individual parts (fuselage, wing, etc.) and by calculating the inertia properties of each part [7]. The longitudinal aerodynamic characteristics of the UC^2AV that were used in the simulation validation can be seen in Eq. (5.6). These values were derived through wind tunnel and flight testing.

$$
\begin{aligned}
C_{L_{0_1}} &= 0.20, \ C_{D_{0_1}} = 0.04, \ C_{L_{\alpha_1}} = 4.41 \\
C_{L_{0_2}} &= 0.55, \ C_{D_{0_2}} = 0.05, \ C_{L_{\alpha_2}} = 4.54 \\
C_{L_{q_1}} &= C_{L_{q_2}} = 1.89 \\
C_{L_{\delta e_1}} &= C_{L_{\delta e_2}} = 0.09
\end{aligned}
\tag{5.6}
$$

The lateral aerodynamic characteristics are shown in Eq. (5.7). The control and stability derivatives appearing in Eq. (4.6) but missing from Eq. (5.7) are equal to 0. The lower bounds were derived by using the flight simulation estimation technique while the upper bounds were given a safe maximum value based on the analysis performed in Sect. 4.2, with any unwanted effects intended to be mitigated by disturbance rejection.

$$
\begin{aligned}
C_{Y_{\beta_1}} &= -0.84, \ C_{Y_{\beta_2}} = 1.5, \ C_{Y_{\delta r_1}} = 0.02 \\
C_{Y_{\delta r_2}} &= 2.5, \ C_{p_{\beta_1}} = -0.17, \ C_{p_{\beta_2}} = 3.13 \\
C_{p_{p_1}} &= -0.42, \ C_{p_{p_2}} = 2.43, \ C_{p_{r_1}} = 0.19 \\
C_{p_{r_2}} &= 2.6, \ C_{p_{\delta\alpha_1}} = 0.13, \ C_{p_{\delta\alpha_2}} = 1.35 \\
C_{p_{\delta r_1}} &= 0.16, \ C_{p_{\delta r_2}} = 1.12
\end{aligned}
\tag{5.7}
$$

The maximum values for the aircraft's physical and control input constraints presented in Eqs. (4.9) are given in Eq. (5.8).

$$
\begin{aligned}
V_{\alpha_{max}} &= 25, \ \rho_{max} = 1.522, \ \alpha_{max} = \pi/4 \\
\gamma_{max} &= \pi/4, \ \beta_{max} = 0.087, \ \theta_{max} = \pi/2 \\
\phi_{max} &= \pi/2, \ \delta_{e_{max}} = \pi/6, \ \delta_{\alpha_{max}} = \pi/3 \\
\delta_{r_{max}} &= \pi/3, \ q_{max} = 0.4, \ p_{max} = 0.8 \\
r_{max} &= 0.8, \ F_{T_{max}} = 45
\end{aligned}
\tag{5.8}
$$

The remaining parameters utilized in the simulation can be seen in Eq. (5.9).

$$e = 0.9$$

$$S = 0.52 \ m^2$$

$$AR = 8.32 \tag{5.9}$$

$$k_{V_\alpha} = 1.5, \ k_\gamma = 1.15, \ k_\beta = 1.44$$

$$\omega_\phi = 1, \ \zeta_\phi = 4.8$$

The additive uncertainty weighting functions for V_α, γ, β and ϕ are given by Eq. (5.10).

$$W_{V_\alpha} = \frac{0.1s + 0.32}{s + 1.5}, \ W_\gamma = \frac{0.35s + 0.89}{s + 1.15}$$

$$\tag{5.10}$$

$$W_\beta = \frac{0.01s + 0.22}{s + 1.44}, \ W_\phi = \frac{0.12s^2 + 0.24s + 0.033}{s^2 + 9.6s + 1}$$

For all tracking errors, the performance weighting function used throughout the simulation validation process is given by Eq. (5.11).

$$W_p = \frac{0.25s + 0.6}{s + 0.006} \tag{5.11}$$

5.4 MATLAB Simulation Results

This section presents MATLAB simulation results for the UC^2AV, derived by running simulation tests with the μ-analysis and synthesis toolbox with no disturbances and only taking into consideration the uncertainty range for each time-varying aerodynamic parameter. The closed loop responses for V_α, γ, β and ϕ using D-K iteration are shown in Figs. 5.22, 5.23, 5.24 and 5.25 respectively. The UC^2AV is commanded to take-off and then perform a circular maneuver with the airspeed reference instruction increasing exponentially up to 21 m/s. For the flight path angle, a step reference instruction is given, occurring at 5 s with an amplitude of 0.26 rad. For roll angle, the UC^2AV is commanded to turn following a step reference instruction that occurs at 12 s with an amplitude of 0.5 rad. Throughout the maneuver (35 s), sideslip angle is required to stay close to 0 rad for lateral stability. The UC^2AV closed loop responses to the reference instructions in all channels are almost ideal (Fig. 5.21).

The UC^2AV D-K iteration summary can be seen in Fig. 5.21. The controller is of order 26, requiring 2 iterations of D-K optimization. The control effort for thrust, elevator, rudder and aileron that was required to perform this maneuver can be seen in Figs. 5.26, 5.27, 5.28 and 5.29 respectively.

DK Iteration Summary				
Iteration #	1	2		
Total D Order	0	8		
Controller Order	18	26		
Gamma Achieved	1.4606	1.0717		
Peak Mu Value	1.4497	0.857		
	<<<		>>>	

Fig. 5.21   D-K iteration summary for the UC^2AV

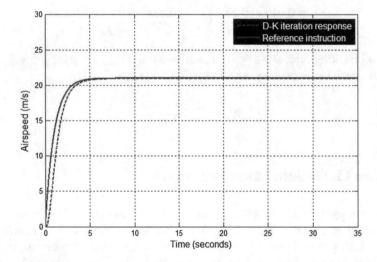

Fig. 5.22   Airspeed response using D-K iteration

5.5   X-Plane Simulation Results

The proposed controller design methodology has been extensively tested on X-Plane flight simulator by performing three benchmark maneuvers, take-off and cruise, take-off and circular and take-off and spiral. Throughout the simulations, five wind conditions were considered and they are depicted in Table 5.1. Specifically, the wind conditions (a–e) range from no wind to strong wind of 13.3 mph, with the average being 6–12 mph according to https://sciencing.com/average-daily-wind-speed-24011.html.

Wind hits the aircraft following the gust and turbulence framework described in Sect. 4.6 and the time intervals this is happening is highlighted in Table 5.1. In what

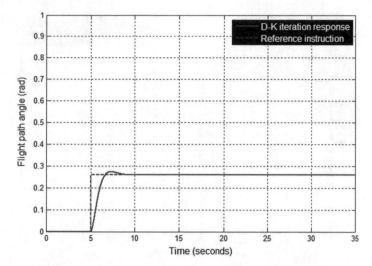

Fig. 5.23 Flight path angle response using D-K iteration

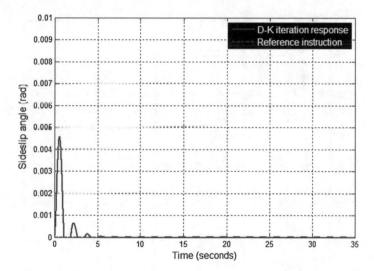

Fig. 5.24 Sideslip angle response using D-K iteration

follows, UC^2AV SIL responses for airspeed, flight path angle, sideslip angle and roll angle are illustrated for each maneuver and each wind condition mentioned above.

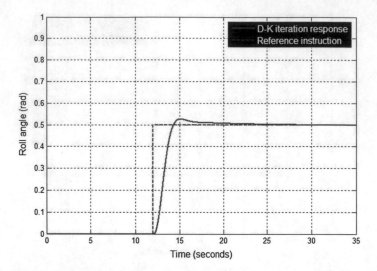

Fig. 5.25 Roll angle response using D-K iteration

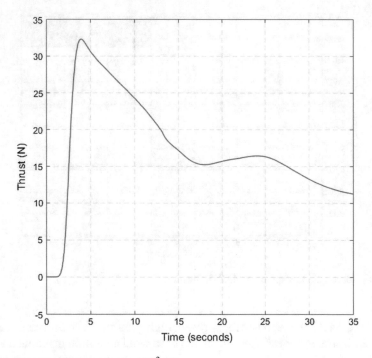

Fig. 5.26 Commanded thrust for the UC^2AV

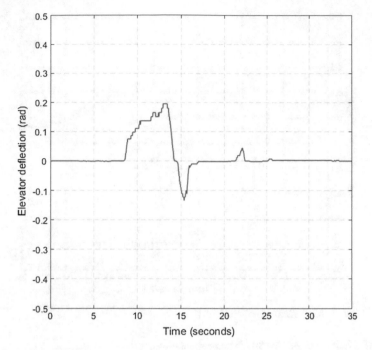

Fig. 5.27 Commanded elevator deflection for the UC^2AV

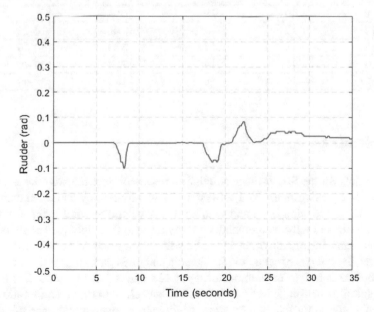

Fig. 5.28 Commanded rudder for the UC^2AV

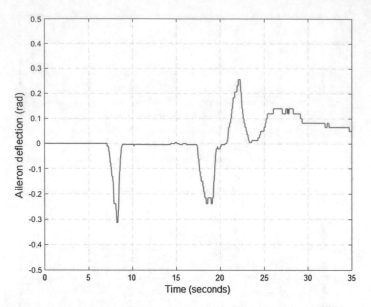

Fig. 5.29   Commanded aileron for the UC^2AV

Table 5.1   Wind conditions for SIL X-Plane simulations

	w_n	w_e	w_d	V_w (mph)	Time interval (s)
a	0	0	0	0	–
b	2	2	2	3.46	10–20
c	3	4	3	5.83	10–20
d	4	5	4	7.54	30–40
e	8	7	8	13.3	30–40

5.6   Remarks

Wind models for the SIL X-Plane simulation results of Sect. 5.5 were implemented in Simulink by using an enabled subsystem, with a time array condition consisting of ones where the wind is on and with zeros for when the wind is off. For cruise flight, sideslip and roll are commanded to stay close to 0°. Flight path angle is given a step reference instruction occurring at 5 s with a magnitude of 14°. For the circular maneuver, the reference instruction for sideslip and flight path angle is the same while roll reference is a step function occurring at 12 s with a magnitude of 28°. Finally, for the spiral maneuver, sideslip is required to stay close to 0, roll angle follows the same reference as for the circular while flight path is given a reference of 30°, for the aircraft to ascend. Airspeed reference is an exponential up to 15 m/s for all cases and maneuvers.

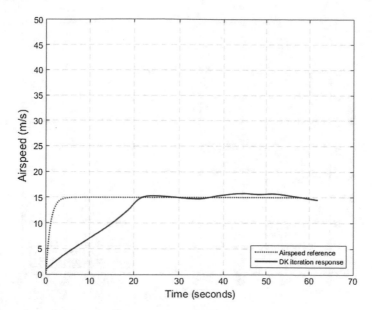

Fig. 5.30 Airspeed response for cruise with wind conditions *a*

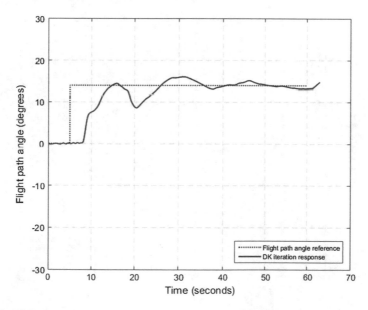

Fig. 5.31 Flight path angle response for cruise with wind conditions *a*

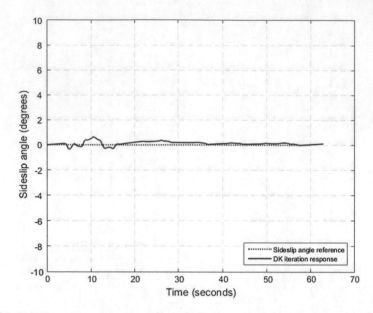

Fig. 5.32 Sideslip angle response for cruise with wind conditions *a*

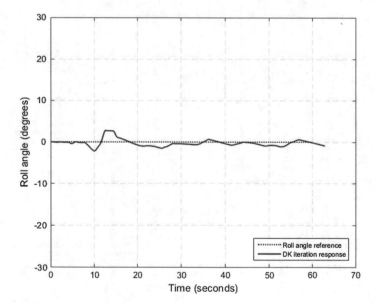

Fig. 5.33 Roll angle response for cruise with wind conditions *a*

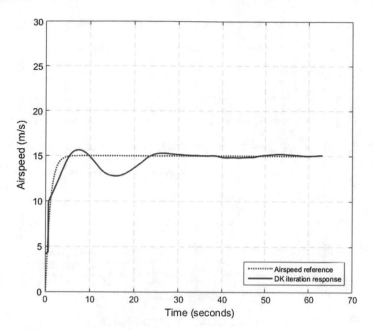

Fig. 5.34 Airspeed response for cruise with wind conditions *b*

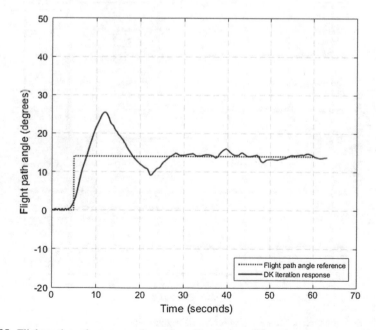

Fig. 5.35 Flight path angle response for cruise with wind conditions *b*

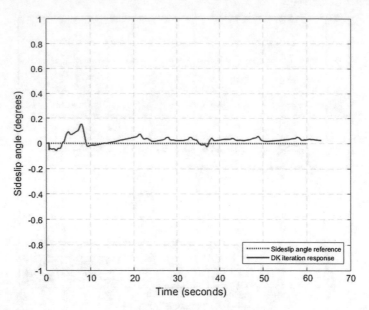

Fig. 5.36 Sideslip angle response for cruise with wind conditions *b*

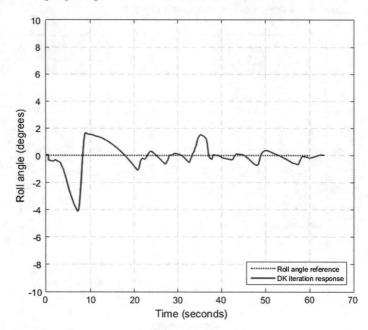

Fig. 5.37 Roll angle response for cruise with wind conditions *b*

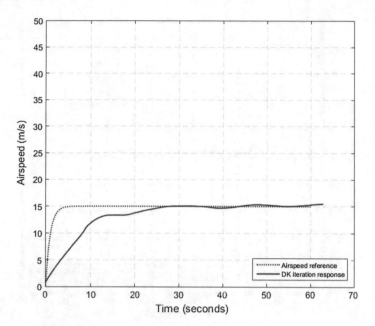

Fig. 5.38 Airspeed response for cruise with wind conditions c

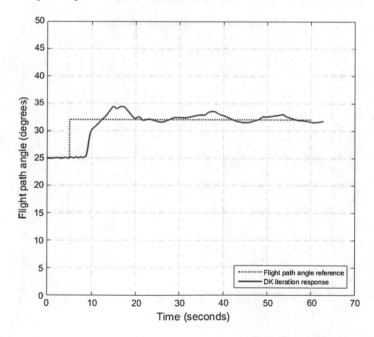

Fig. 5.39 Flight path angle response for cruise with wind conditions c

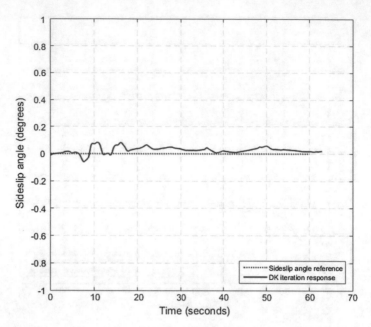

Fig. 5.40 Sideslip angle response for cruise with wind conditions c

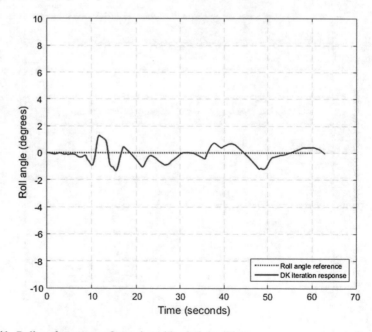

Fig. 5.41 Roll angle response for cruise with wind conditions c

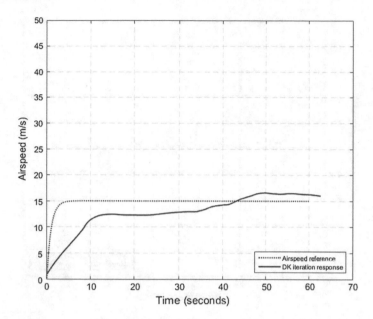

Fig. 5.42 Airspeed response for cruise with wind conditions d

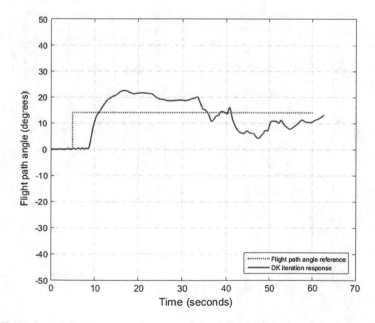

Fig. 5.43 Flight path angle response for cruise with wind conditions d

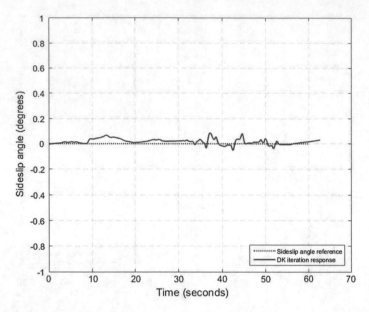

Fig. 5.44   Sideslip angle response for cruise with wind conditions d

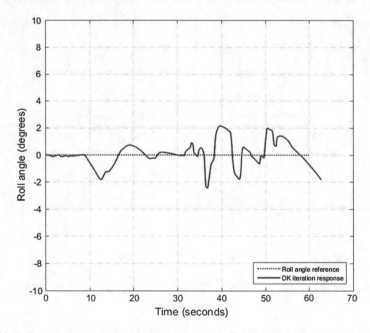

Fig. 5.45   Roll angle response for cruise with wind conditions d

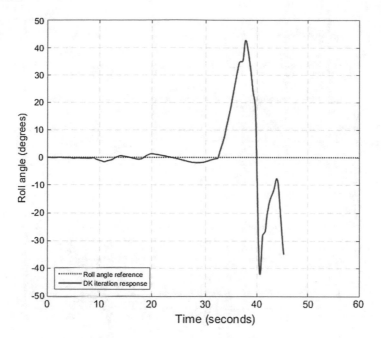

Fig. 5.46 Roll angle response for cruise with wind conditions *e*

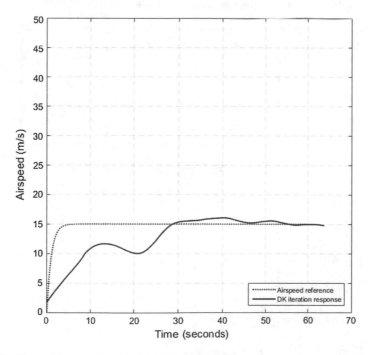

Fig. 5.47 Airspeed response for circular with wind conditions *a*

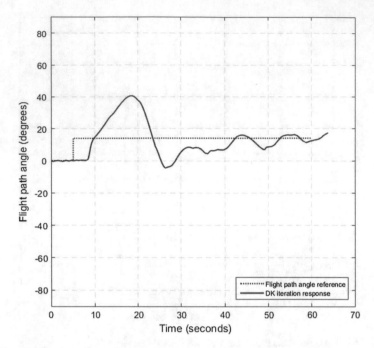

Fig. 5.48   Flight path angle response for circular with wind conditions *a*

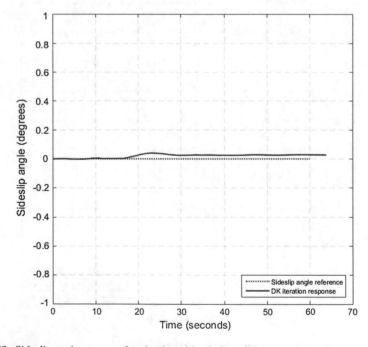

Fig. 5.49   Sideslip angle response for circular with wind conditions *a*

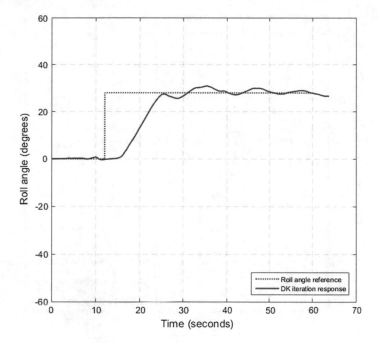

Fig. 5.50 Roll angle response for circular with wind conditions *a*

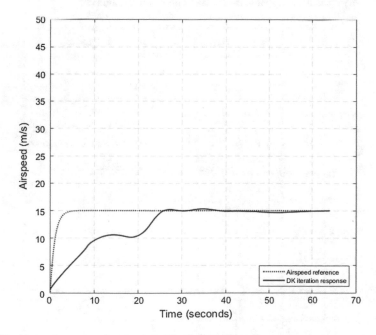

Fig. 5.51 Airspeed response for circular with wind conditions *b*

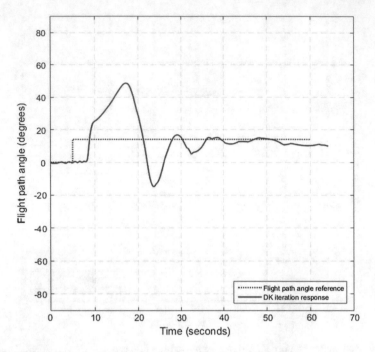

Fig. 5.52 Flight path angle response for circular with wind conditions b

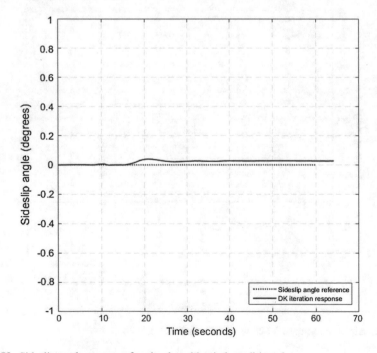

Fig. 5.53 Sideslip angle response for circular with wind conditions b

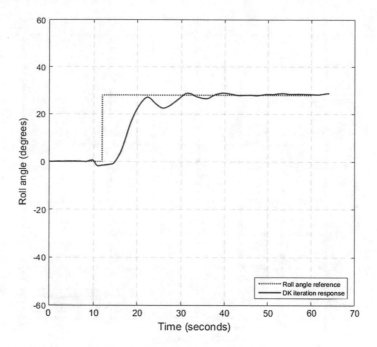

Fig. 5.54 Roll angle response for circular with wind conditions *b*

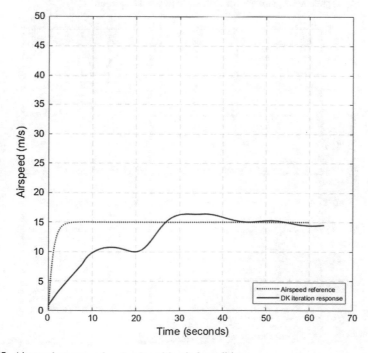

Fig. 5.55 Airspeed response for circular with wind conditions *c*

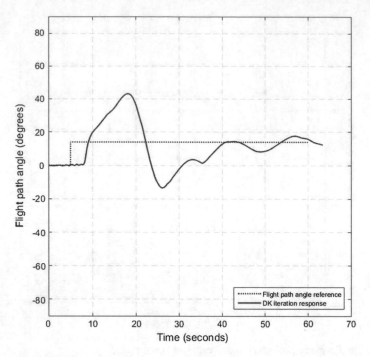

Fig. 5.56 Flight path angle response for circular with wind conditions c

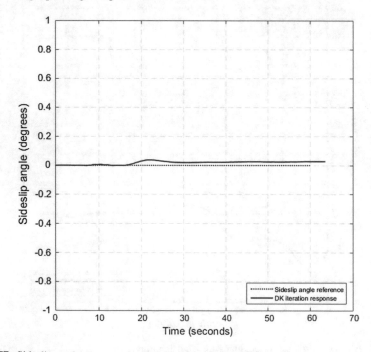

Fig. 5.57 Sideslip angle response for circular with wind conditions c

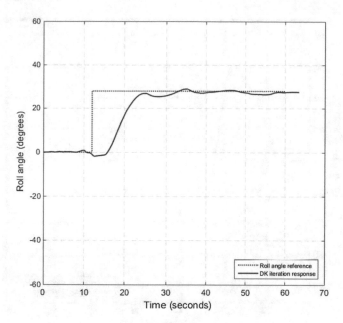

Fig. 5.58 Roll angle response for circular with wind conditions *c*

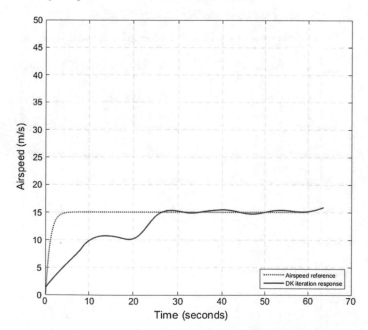

Fig. 5.59 Airspeed response for circular with wind conditions *d*

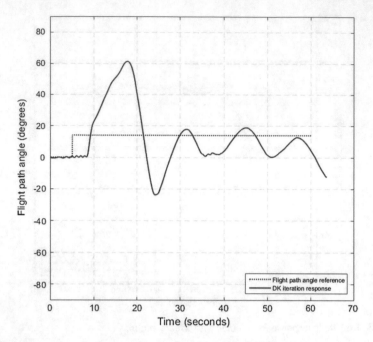

Fig. 5.60 Flight path angle response for circular with wind conditions d

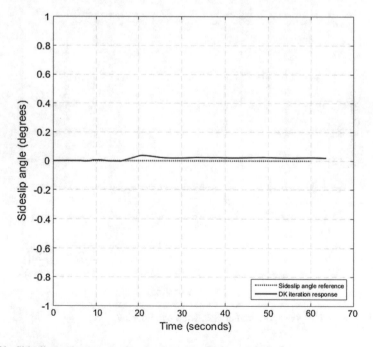

Fig. 5.61 Sideslip angle response for circular with wind conditions d

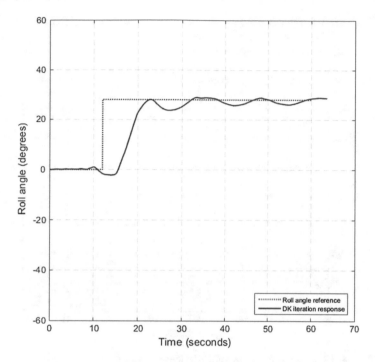

Fig. 5.62 Roll angle response for circular with wind conditions *d*

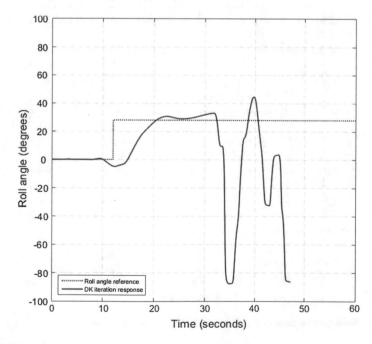

Fig. 5.63 Roll angle response for circular with wind conditions *e*

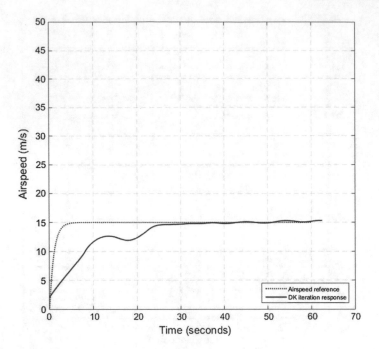

Fig. 5.64 Airspeed response for spiral with wind conditions a

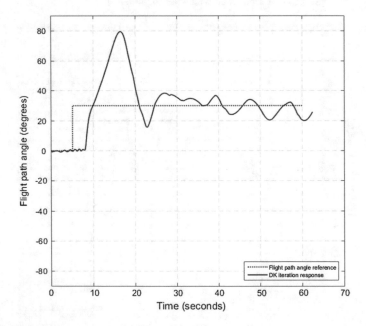

Fig. 5.65 Flight path angle response for spiral with wind conditions a

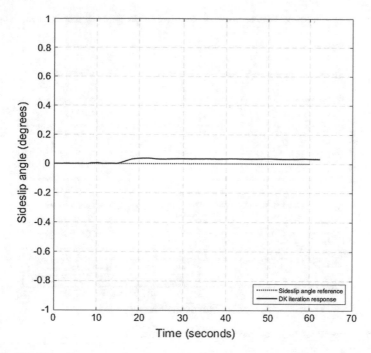

Fig. 5.66 Sideslip angle response for spiral with wind conditions *a*

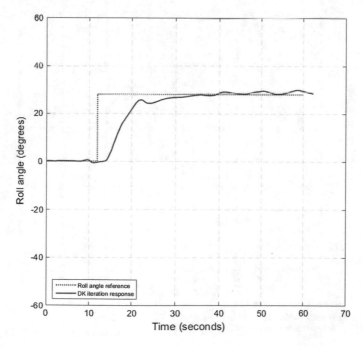

Fig. 5.67 Roll angle response for spiral with wind conditions *a*

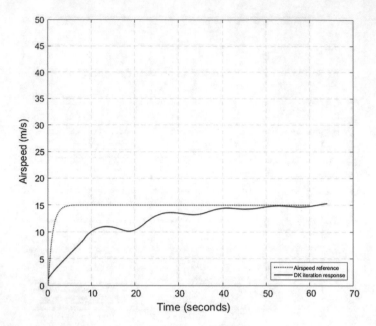

Fig. 5.68 Airspeed response for spiral with wind conditions b

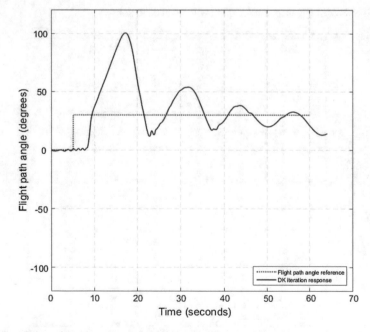

Fig. 5.69 Flight path angle response for spiral with wind conditions b

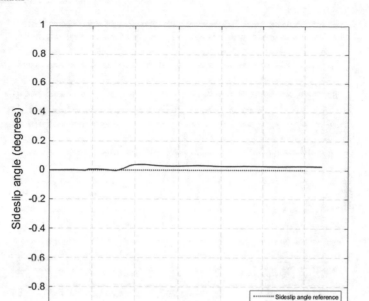

Fig. 5.70 Sideslip angle response for spiral with wind conditions *b*

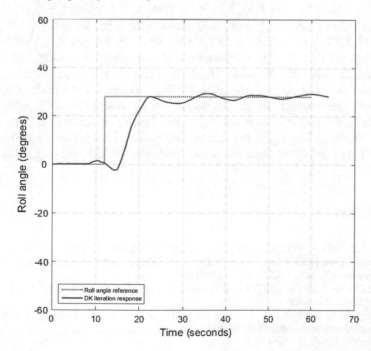

Fig. 5.71 Roll angle response for spiral with wind conditions *b*

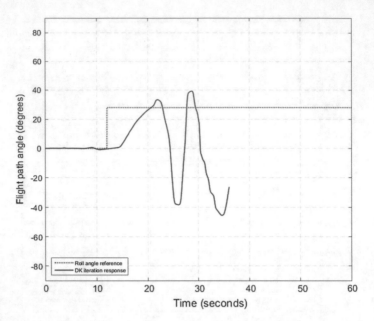

Fig. 5.72 Roll angle response for spiral with wind conditions c

Figures 5.30, 5.31, 5.32, 5.33, 5.34, 5.35, 5.36, 5.37, 5.38, 5.39, 5.40, 5.41, 5.42, 5.43, 5.44, 5.45, 5.46, 5.47, 5.48, 5.49, 5.50, 5.51, 5.52, 5.53, 5.54, 5.55, 5.56, 5.57, 5.58, 5.59, 5.60, 5.61, 5.62, 5.63, 5.64, 5.65, 5.66, 5.67, 5.68, 5.69, 5.70, 5.71 and 5.72 prove that the proposed technique is promising and provides acceptable efficiency. The aircraft's responses are almost ideal in most cases given a realistic gust model and wind variety. Aircraft's stability was tested to the limit to highlight wind disturbance bounds for current design tuning parameters and specifications. Specifically, the UC^2AV loses stability in three cases; (i) cruise flight with a total wind of 13.3 mph (Fig. 5.46), (ii) circular maneuver with a total wind of 13.3 mph (Fig. 5.63) and (iii) spiral maneuver with a total wind of 5.83 mph (Fig. 5.72).

References

1. Michailidis MG, Agha M, Kanistras K, Rutherford MJ, Valavanis KP (2017) A controller design framework for a NextGen circulation control based UAV. In: IEEE conference on control technology and applications (CCTA), pp 1542–1549
2. Beard RW, McLain TW (2012) Small unmanned aircraft: theory and practice. Princeton University Press, Princeton
3. Kanistras K, Rutherford MJ, Valavanis KP (2018) Foundations of circulation control based small-scale unmanned aircraft. Springer, Berlin
4. Stevens BL, Lewis FL, Johnson EN (2015) Aircraft control and simulation: dynamics, controls design, and autonomous systems. Wiley, Hoboken

5. Michailidis MG, Kanistras K, Agha M, Rutherford MJ, Valavanis KP (2017) Robust nonlinear control of the longitudinal flight dynamics of a circulation control fixed wing UAV. In: IEEE conference on decision and control (CDC), pp 3920–3927
6. Michailidis MG, Kanistras K, Agha M, Rutherford MJ, Valavanis KP (2018) Nonlinear control of fixed-wing UAVs with time-varying aerodynamic uncertainties via μ-synthesis. In: IEEE conference on decision and control (CDC), pp 6314–6321
7. Klein V, Morelli EA (2006) Aircraft system identification: theory and practice. American Institute of Aeronautics and Astronautics, Reston

Chapter 6
Conclusion and Discussion

Abstract This chapter concludes the monograph, re-stating the major contributions and results. Potential future research directions that may extend applicability of the methodology for robust control of an arbitrary nonlinear system are also discussed.

This research monograph presented a hybrid, nonlinear, inner-outer loop autopilot for fixed wing UAVs with time-varying aerodynamic parameters, consisting of a dynamic inversion and a μ-synthesis controller. The novelty the author is proposing relates to the uncertainty range of the respective, uncertain parameters with time-varying structure and an appropriate modification of unmodeled dynamics via additive uncertainty weighting functions. MATLAB and software-in-the-loop, Simulink/XPlane results for a new generation UAV (UC^2AV) were demonstrated, proving that the proposed technique is reliable.

The focal point that may limit applicability of the proposed nonlinear controller design to any type of nonlinear system, from ground vehicles to spacecraft, is the uncertainty range itself. In the ideal case, system identification will be available for a given system so the uncertainty range will be known and utilized for the derivation of the uncertainty weighting functions. Analysis that was conducted herein shows that with extreme uncertainty bounds and appropriately tuned performance weighting functions, robust performance can be achieved following the proposed methodology. The question is how can one optimally adjust the uncertainty interval for a certain system with specific time-varying uncertainties.

Is it possible to have a stochastic, multi-model, adaptation algorithm that somehow adjusts the lower/upper bounds accordingly for random uncertainties and any nonlinear system? Work in this direction was initially proposed in [1], an overview of which is shown in Fig. 6.1. Local-non adaptive compensators are blended with multi-estimators achieving robust performance based on a multi-level Kalman filtering and system identification.

As a result, the proposed nonlinear controller framework can be integrated with a multi-model algorithm for system identification, thus making the technique a rigid and mathematically rigorous method, applicable to any nonlinear system.

An alternative approach is an optimal adjustment of the uncertainty intervals by employing machine learning techniques to tackle challenges stemming from system identification. Machine learning is already being used for enabling adaptive

© Springer Nature Switzerland AG 2020   117
M. G. Michailidis et al., *Nonlinear Control of Fixed-Wing UAVs
with Time-Varying and Unstructured Uncertainties*, Springer Tracts
in Autonomous Systems 1, https://doi.org/10.1007/978-3-030-40716-2_6

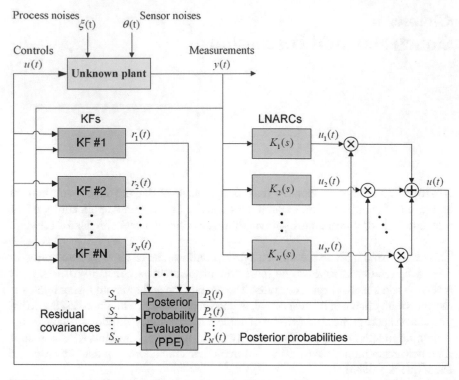

Fig. 6.1 Robust multi-model adaptive architecture [1]

autonomy. Machine learning is grouped in supervised, unsupervised, or reinforcement, depending on the amount and quality of feedback about the system or task [2–4]. In supervised learning, the feedback information provided to learning algorithms is a labeled training data set, and the objective is to build the system model representing the learned relation between the input, output, and system parameters. In unsupervised learning, no feedback information is provided to the algorithm and the objective is to classify the sample sets to different groups based on the similarity between the input samples. Finally, reinforcement learning (RL) is a goal-oriented learning tool wherein the agent or decision maker learns a policy to optimize a long-term reward by interacting with the environment. At each step, an RL agent gets evaluative feedback about the performance of its action, allowing it to improve the performance of subsequent actions.

The goal in a problem like this is to learn the optimal policy and value function for a potentially uncertain system, with time-varying uncertainties. Unlike traditional optimal control, RL finds the solution to the HJB equation online in real time. Recent applications of iterative and reinforcement learning can be seen in [5–7] for robotic arm and UAV attitude control respectively.

Machine learning has been accepted as a great enabler to understand scientific and engineering data. The use of machine learning to design controllers of physical systems is, however, still in its infancy. Relevant work includes the one on discovery

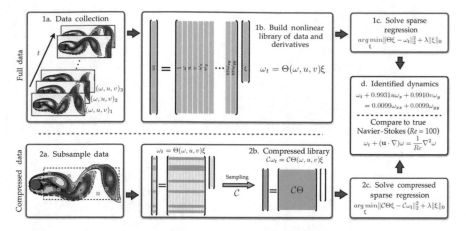

Fig. 6.2 Framework of data driven learning of nonlinear dynamics [8]

of physical laws governing the behavior of a system from data [8]. Fast methods can be developed to learn coarse models on the fly, that satisfy control relevant properties such as passivity that allow fast real-time controller design. Also relevant is the work on reinforcement learning. In particular, ways can be obtained to optimally adjust the uncertainty interval on the model parameters and terms, and learn these values using data-driven techniques continuously (Fig. 6.2). Using regression methods and sparsity techniques, and by enforcing properties such as dissipativity that we know a priori that the system has to satisfy, will enable us to learn a control-relevant model that accurately represents collected data and is adapted continuously.

References

1. Fekri S, Athans M, Pascoal A (2006) Issues, progress and new results in robust adaptive control. Int J Adapt Control Signal Process 20(10):519–579
2. Kiumarsi B, Vamvoudakis KG, Modares H, Lewis FL (2018) Optimal and autonomous control using reinforcement learning: a survey. IEEE Trans Neural Netw Learn Syst 29(6):2042–2062
3. He W, Meng T, He X, Ge SS (2018) Unified iterative learning control for flexible structures with input constraints. Automatica 96:326–336
4. Mars P (2018) Learning algorithms: theory and applications in signal processing, control and communications. CRC Press, Boca Raton
5. Meng T, He W (2018) Iterative learning control of a robotic arm experiment platform with input constraint. IEEE Trans Ind Electron 65(1):664–672
6. Choromanski K, Sindhwani V, Jones B, Jourdan D, Chociej M, Boots B (2018) Learning-based air data system for safe and efficient control of fixed-wing aerial vehicles. In: IEEE international symposium on safety, security, and rescue robotics (SSRR)
7. Koch W, Mancuso R, West R, Bestavros A (2019) Reinforcement learning for UAV attitude control. ACM Trans Cyber-Phys Syst
8. Rudy SH, Brunton SL, Proctor JL, Kutz JN (2017) Data-driven discovery of partial differential equations. Sci Adv